New York State Regents Examination Coach, Geometry

Coach™
America's Best for Student Success®

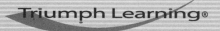
Triumph Learning®

Jerome D. Kaplan, Ed.D.
Senior Math Advisor

New York State Regents Examination Coach, Geometry
212NY
ISBN-10:1-59823-756-X
ISBN-13:978-1-59823-756-6

Cover Image: The apple is the state fruit of New York. ©JupiterImages

Triumph Learning® 136 Madison Avenue, 7th Floor, New York, NY 10016
Kevin McAliley, President and Chief Executive Officer

Printed in the United States of America.

10 9 8 7 6 5 4 3 2 1

Table of Contents

			New York State Indicators
Chapter 1	**Introduction to Informal and Formal Proofs**		
	Lesson 1	Statements, Negations, and Truth Values. 8	G.G.24, G.G.25
	Lesson 2	Conditionals and Biconditionals. 14	G.G.25
	Lesson 3	Inverses, Converses, and Contrapositives 19	G.G.26
	Lesson 4	The Laws of Logic 24	G.G.27
	Lesson 5	Logic Proofs: Hypothesis to Conclusion 32	G.G.27
Chapter 2	**Geometric Relationships**		
	Lesson 6	Inductive and Deductive Reasoning. 39	G.G.27
	Lesson 7	Basic Postulates. 44	G.G.27
	Lesson 8	Geometric Definitions and Postulates 51	G.G.27
	Lesson 9	Perpendicular Lines and Planes. 61	G.G.1, G.G.2, G.G.3, G.G.4, G.G.5, G.G.6, G.G.7
	Lesson 10	Parallel Lines and Planes 67	G.G.8, G.G.9, G.G.35
Chapter 3	**Three-Dimensional Figures**		
	Lesson 11	Prisms. 75	G.G.10, G.G.11, G.G.12
	Lesson 12	Regular Pyramids 82	G.G.13
	Lesson 13	Cylinders. 87	G.G.14
	Lesson 14	Cones 93	G.G.15
	Lesson 15	Spheres. 99	G.G.16

Letter to the Student

Dear Student,

Welcome to the **New York State Regents Examination Coach, Geometry**. This book will help you to strengthen your mathematics skills this year. Coach also provides practice with the types of questions you will have to answer on tests, including the state test.

The *Coach* book is divided into chapters and lessons. Before you begin the first chapter, you may want to take the Practice Test 1. The Practice Test 1 will show you your strengths and weaknesses in the skills and strategies you need to know this year. This way, you will be aware of what you need to concentrate on to be successful. At the end of the *Coach* book is a Practice Test 2 that will allow you and your teacher to evaluate how much you have learned.

The lessons in this book will help you review and practice your skills and get you ready to take tests. Some of the practice will be in the style of the state test. In general, you will be answering multiple choice and open-ended questions. Questions like these may appear on your state test. Practicing with these types of questions will give you a good idea of what you need to review to triumph.

Here are some tips that will help you as you work through this book. Remembering these tips will also help you do well on the state test.

- Listen closely to your teacher's directions.
- When answering multiple-choice questions, read each choice carefully before choosing the BEST answer.
- When answering open-ended questions, think about how you will answer the question before you begin to write.
- Time yourself so that you have the time at the end of a test to check your answers.

We hope you will enjoy using *Coach* and that you will have a fun and rewarding year!

Letter to the Family

Dear Parents and Families,

The *Coach* series of workbooks is designed to prepare your child to master grade-appropriate skills in mathematics and to take the Geometry Regents Examination, which is the test administered each year in the state of New York. In your state, the grade-appropriate skills are called Indicators. These are the skills the state has chosen as the building blocks of your child's education in mathematics, and these are the skills that will be tested on the Geometry Regents Examination. Your child's success will be measured by how well he or she masters these skills.

You are an important factor in your child's ability to learn and succeed. Get involved! We invite you to be our partner in making learning a priority in your child's life. To help ensure success, we suggest that you review the lessons in this book with your child. While teachers will guide your child through the book in class, your support at home is also vital to your child's comprehension.

Please encourage your child to read and study this book at home, and take the time to go over the sample questions and homework together. The more students practice, the better they do on the actual exam and on all the tests they will take in school. Try talking about what your child has learned in school. Perhaps you can show your children real-life applications of what they have learned. For example, you could discuss why math skills are important in life and how they apply to everyday situations.

You will also want to foster good study habits. Students should set aside quiet time every day to do their homework and study for tests. Children need to learn to pace themselves to avoid cramming, or last minute preparing, for the challenges they will face in school. These are behaviors that young students carry with them for life.

We ask you to work with us this year to help your child triumph. Together, we can make a difference!

Statements, Negations, and Truth Values

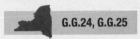
G.G.24, G.G.25

A **statement** is a mathematical sentence that can be proven true or false. When you determine whether a statement is true or false, you determine its **truth value**. "A rectangle has four sides" is a true statement. "10 + 3 = 12" is a false statement. To simplify the process of determining truth values, statements are written in symbolic notation. The letters p, q, and r are typically used to represent statements or parts of statements.

The **negation** of a statement is the opposite of the original statement. The statement is often converted to a negation by adding or removing the word "not" from the statement. The symbol $\sim$ is used to show the negation of a statement and $\sim p$ is read "not p".

Statement	Symbol	Negation	Symbol
I will go to school.	p	I will not go to school.	$\sim p$
I am not sick.	q	I am sick	$\sim q$

EXAMPLE 1

Complete the table by inserting the correct statement, negation, and symbol. Give the truth value of the original and the negation.

Statement	Symbol	Negation	Symbol
A line segment does not have two endpoints.	p		

STRATEGY   **Use the definition of negation.**

STEP 1   Write the negation of the statement as the opposite of the statement. Because the original statement contains the word "not", write the negation by removing this word. The negation of the statement is "a line segment has two endpoints."

STEP 2   Write the symbol for the negation. If the original statement is symbolized by p, then the negation uses the symbol $\sim p$.

STEP 3   Complete the table and determine if the original statement or negation is true. Since a segment has two endpoints, the original statement is false and the negation is true.

SOLUTION   **In this example, the original statement is false and the negation is true.**

Statement	Symbol	Negation	Symbol
A line segment does not have two endpoints.	p	A line segment has two endpoints.	$\sim p$

A **conjunction** is a compound sentence combining two statements with the word "and." A conjunction is only true if both statements are true. The symbol $\wedge$ represents the word "and" in symbolic notation. Remember the symbol by thinking that it is similar in shape to the letter A in the word "And". A **truth table** shows all of the possible truth values of a compound statement.

EXAMPLE 2

Given the true statement "Alicia plays soccer" and the false conjunction "Alicia plays soccer and Alicia plays piano", create a truth table to determine the truth value of the statement "Alicia plays piano."

STRATEGY   Create a truth table.

STEP 1   Let p represent the statement "Alicia plays soccer" and let q represent the statement "Alicia plays piano." Then $p \wedge q$ represents the conjunction "Alicia plays soccer and Alicia plays piano."

STEP 2   Create a truth table that represents all possible combinations of truth values for each statement.

p	q	$p \wedge q$
T	T	T
T	F	F
F	T	F
F	F	F

STEP 3   Locate the row in the truth table where p is true and $p \wedge q$ is false. The truth value of q in this row will be the solution to the problem.

p	q	$p \wedge q$
T	T	T
T	F	F
F	T	F
F	F	F

SOLUTION   The truth value of the statement "Alicia plays piano" is false.

A **disjunction** is a compound sentence combining two statements with the word "or". A disjunction is true if one or both statements are true. The symbol $\vee$ represents the word "or" in symbolic notation.

p	q	$p \vee q$
T	T	T
T	F	T
F	T	T
F	F	F

EXAMPLE 3

Given the true statement "Tyrese will go to the movies or Tyrese will wash his car" and the false statement "Tyrese will wash his car", determine the truth value of the statement "Tyrese will go to the movies."

STRATEGY   **Use the definition of disjunction.**

Because the disjunction "Tyrese will go to the movies or Tyrese will wash his car" is true, then by definition of a disjunction, one or both of the statements in the disjunction must be true. It is given that the statement "Tyrese will wash his car" is false. Therefore, the other statement must be true.

SOLUTION   **The truth value of the statement "Tyrese will go to the movies" is true.**

COACHED EXAMPLE

Given the true statement "Jeremy lives in Albany and Jeremy lives in New York State", determine the truth value of the statement "Jeremy lives in Albany."

THINKING IT THROUGH

The statement "Jeremy lives in Albany and Jeremy lives in New York" is a _____ because it links two statements using the word _____.

In order for this statement to be true, the statement "Jeremy lives in Albany" must be _____ and the statement "Jeremy lives in New York State" must be _____.

Therefore, the truth value of the statement "Jeremy lives in Albany" is _____.

Lesson Practice

Choose the correct answer.

1. Which of the following is the negation of the statement "The sum of the angles of a triangle is 180°"?

   (1) The sum of the angles of a square is 360°.

   (2) The sum of the angles of a triangle is not 180°.

   (3) A triangle has three sides.

   (4) The sum of the angles of a triangle is 180°, and a triangle has three sides.

2. Which of the following is a disjunction?

   (1) Angela is forty-three.

   (2) Lisa will run a marathon and Lisa will compete in a triathlon.

   (3) Paul is at school or Paul is at home.

   (4) A square has four sides and four angles.

3. Let p represent the statement "The square root of 49 is 7." Which of the following is $\sim p$?

   (1) The square root of 49 is not 7.

   (2) The square root of 49 may or may not be 7.

   (3) The square of 7 is 49.

   (4) The square root of 64 is 8.

4. Which of the following correctly shows the truth table for a conjunction?

   (1)

p	q	$p \wedge q$
T	T	T
T	F	F
F	T	F
F	F	F

   (2)

p	q	$p \wedge q$
T	T	T
T	F	F
F	T	T
F	F	T

   (3)

p	q	$p \wedge q$
T	T	T
T	F	T
F	T	T
F	F	F

   (4)

p	q	$p \wedge q$
T	T	T
T	F	T
F	T	F
F	F	F

5. Given the true statement "Kyle is a painter or Kyle is a musician" and the true statement "Kyle is a painter", which of the following is true?

   (1) Kyle is not a musician.

   (2) "Kyle is a musician" could have either a true or a false truth value.

   (3) "Kyle is a painter and a musician" can only have a false truth value.

   (4) "Kyle is a musician" can only have a true truth value.

6. Given the true statement "Brianna is majoring in physics and Brianna attends Stony Brook University", which of the following is true?

   (1) "Brianna attends Stony Brook University" may or may not be a true statement.

   (2) "Brianna is majoring in Physics" must be a false statement.

   (3) "Brianna attends Stony Brook University" must be a true statement.

   (4) "Brianna is majoring in Physics" may or may not be a true statement.

7. Which of the following is a conjunction?

   (1) A stop sign is an octagon or a stop sign is not an octagon.

   (2) There is road construction and the traffic is delayed.

   (3) $\frac{10}{2} = 5$ or $5 \times 2 = 10$.

   (4) Eileen is a doctor or a dentist.

8. Given the false statement "The dog is a Chihuahua or the dog is brown" and the false statement "The dog is a Chihuahua", which of the following is true?

   (1) The truth value of the statement "the dog is a Chihuahua" is true.

   (2) The truth value of the statement "the dog is brown" is true.

   (3) The truth value of the statement "the dog is brown" is false.

   (4) The truth value of the statement "the dog is brown" could be true or false.

OPEN-ENDED QUESTIONS

9. Given the statement: "Jose received a speeding ticket or Jose's insurance rate increased."

    A. Create a truth table representing all truth values for p, q, and $p \lor q$.

    B. If q is false, and $p \lor q$ is true, use the truth table to find the truth value of p.

    _____

10. Given the statement: "The supplement of a 100° angle is a 90° angle."

    A. Write the negation of this statement.

    _____

    B. Which is true, the statement or the negation?

    _____

Conditionals and Biconditionals

 G.G.25

A **conditional statement** is a statement of the form "if-then". The **hypothesis** is the part of the statement that follows "if". It is a proposal of certain conditions. The **conclusion** is the part of the statement that follows "then". It is the result of the hypothesis being true. In a truth table, the hypothesis is denoted by p and the conclusion is denoted by q. The statement "if p then q" is written $p \rightarrow q$. Any statement can be put in conditional form.

EXAMPLE 1

Identify the hypothesis and conclusion of the following statement.
The fastest land bird is an ostrich.

STRATEGY   **Write the statement in if-then form.**

STEP 1   Write the statement as a conditional in if-then form.

If a bird is the fastest land bird, then it is an ostrich.

STEP 2   Identify the hypothesis.

The hypothesis follows the word "if". The hypothesis is "If a bird is the fastest land bird."

STEP 3   Identify the conclusion.

The conclusion follows the word "then". The conclusion is "then it is an ostrich."

SOLUTION   **The hypothesis is "If a bird is the fastest land bird". The conclusion is "then it is an ostrich".**

EXAMPLE 2

Create a truth table for the conditional statement: If the video game costs less than $25, then Paul will buy it.

STRATEGY **Find all possible truth values of the statement.**

STEP 1 Let *p* represent the hypothesis: The video game costs less than $25.
Let *q* represent the conclusion: Paul will buy it.

STEP 2 Let both *p* and *q* be true: If the video game costs less than $25, then Paul will buy it. This is true. Therefore, $p \rightarrow q$ is true.

STEP 3 Let *p* be true and *q* be false: If the video game costs less than $25, then Paul will not buy it. According to the original statement, this is false. Therefore, $p \rightarrow q$ is false.

STEP 4 Let *p* be false; this means the hypothesis is: The video game does not cost less than $25. Because we cannot conclude what Paul will do if the video game costs less than $25, we assign a true value to this statement, whether *q* is true or false. This is a convention in the field of logical proofs. Therefore, $p \rightarrow q$ is true. Notice that a conditional statement can only be false if the hypothesis is true and the conclusion is false.

STEP 5 Create a truth table of all truth values. This truth table will be the same for all conditional statements.

SOLUTION **The truth table is shown below.**

p	*q*	$p \rightarrow q$
T	T	T
T	F	F
F	T	T
F	F	T

A **biconditional statement** is a compound sentence combining two facts with the words "if and only if." A biconditional statement is only true when both statements are true or both statements are false. The statement "p if and only if q" is written $p \leftrightarrow q$.

p	q	$p \leftrightarrow q$
T	T	T
T	F	F
F	T	F
F	F	T

"An angle is a right angle if and only if it measures 90°" is a biconditional statement. This means both of the following statements are true: An angle is a right angle if it measures 90°; and if an angle measures 90°, then it is a right angle. Definitions are examples of biconditionals.

EXAMPLE 3

Write this statement as a biconditional statement: Two segments with the same length are defined to be congruent.

STRATEGY   **Write the statement in "if and only if" form.**

Let the first statement be: Two segments are congruent.

Let the second statement be: They have the same length.

SOLUTION   **Two segments are congruent if and only if they have the same length.**

COACHED EXAMPLE

Write this definition as a biconditional statement: An equilateral triangle has three congruent sides. Then write two conditional statements that make up the biconditional statement.

THINKING IT THROUGH

A biconditional statement joins two statements with the words _____.

Let one statement be: A triangle is _____.

Let the other statement be: It has _____.

Then the biconditional statement is: _____

_____.

A conditional statement joins a _____ and a _____ with the words "if" and "then".

Let the first statement be the hypothesis and the second statement be the conclusion. If a triangle is _____ then _____

_____.

Now let the second statement be the hypothesis and the first statement be the conclusion.

If _____ then it is _____.

Lesson Practice

Choose the correct answer.

1. Which of the following is an example of a biconditional statement?

   (1) You will win the race if and only if you have the fastest time.

   (2) If you have the fastest time, then you will win the race.

   (3) The winner of the race has the fastest time.

   (4) If you do not have the fastest time, then you will not win the race.

2. Given the true conditional: "If Tai attends all of the practices, then he will start in Sunday's game." Which of the following does not have a true value?

   (1) Tai attends all of the practices. Tai starts in Sunday's game.

   (2) Tai attends all of the practices. Tai does not start in Sunday's game.

   (3) Tai does not attend all of the practices.

   (4) Tai does not start in Sunday's game.

3. Which biconditional is formed from these two conditionals: "If a number is even, then it is divisible by two" and, "If a number is divisible by two, then it is even"?

   (1) Even numbers are divisible by two.

   (2) Numbers that are divisible by two are even.

   (3) If a number is divisible by two and it is even, then it is a multiple of two.

   (4) A number is even if and only if it is divisible by two.

4. Which of the following correctly shows the truth table for a biconditional?

   (1)

p	q	$p \leftrightarrow q$
T	T	T
T	F	F
F	T	F
F	F	F

   (2)

p	q	$p \leftrightarrow q$
T	T	T
T	F	F
F	T	F
F	F	T

   (3)

p	q	$p \leftrightarrow q$
T	T	T
T	F	T
F	T	T
F	F	F

   (4)

p	q	$p \leftrightarrow q$
T	T	T
T	F	F
F	T	T
F	F	F

5. An octagon is a polygon with eight sides. Which of the following correctly shows this definition as a conditional statement?

(1) If an octagon has eight sides, then it is a polygon.

(2) If a polygon is an octagon, then it has eight sides.

(3) If a polygon has eight sides, then it is not an octagon.

(4) An octagon is a polygon if and only if it has eight sides.

6. Which of the following statements is true about a biconditional statement composed of the facts p and q?

(1) A biconditional statement is only true when p is true.

(2) A biconditional statement is only false when both facts are false.

(3) A biconditional statement is only true when both p and q are true or both p and q are false.

(4) A biconditional statement is only false when q is false.

OPEN-ENDED QUESTION

7. Given the definition: A straight angle has a measure of 180°.

A. Create two conditionals from this definition.

_____

_____

B. Create a biconditional statement from these two conditionals.

_____

Inverses, Converses, and Contrapositives

The **converse** of a conditional statement switches the hypothesis and the conclusion.

The **inverse** of a conditional statement negates both the hypothesis and conclusion, usually by adding or removing the word "not" to both.

The **contrapositive** of a conditional statement switches the hypothesis and conclusion and also negates both. It is a conditional statement consisting of the converse and the inverse of a statement.

EXAMPLE 1

Write the converse, inverse, and contrapositive of the statement below.

If a quadrilateral is a square, then it has four right angles and four congruent sides.

STRATEGY Use the definitions of converse, inverse, and contrapositive.

STEP 1 Write the "if...then" statement. Underline the hypothesis, and circle the conclusion.

If a quadrilateral is a square, then it has four right angles and four congruent sides.

STEP 2 Write the converse of the conditional statement by switching the hypothesis and conclusion.

If a quadrilateral has four right angles and four congruent sides, then it is a square.

STEP 3 Write the inverse of the conditional statement by negating both the hypothesis and the conclusion.

If a quadrilateral is not a square, then it does not have four right angles and four congruent sides.

STEP 4 Write the contrapositive of a conditional statement by switching the hypothesis and conclusion and negating both.

If a quadrilateral does not have four right angles and four congruent sides, then it is not a square.

SOLUTION The converse, inverse, and contrapositive are shown in the steps above.

EXAMPLE 2

A conditional statement "if p then q" is represented as $p \rightarrow q$. Use p's and q's to represent the converse, inverse, and contrapositive.

STRATEGY   **Symbolize statements using p's and q's.**

STEP 1   Write the converse by switching the hypothesis, p, and the conclusion, q.

The converse can be written: $q \rightarrow p$

STEP 2   Write the inverse by negating the hypothesis and conclusion.

The negation of the statement p is symbolized as $\sim p$. Then the inverse can be written: $\sim p \rightarrow \sim q$.

STEP 3   Write the contrapositive by negating the hypothesis and conclusion and switching them.

The contrapositive can be written: $\sim q \rightarrow \sim p$.

SOLUTION   **converse: $q \rightarrow p$; inverse: $\sim p \rightarrow \sim q$; contrapositive: $\sim q \rightarrow \sim p$**

EXAMPLE 3

Write the inverse of the statement below. Determine the possible truth values by creating a truth table.

If two lines intersect, then they are not parallel.

STRATEGY   **Use the definition of inverse.**

STEP 1   Write the inverse of the conditional statement by negating both the hypothesis and the conclusion.

If two lines do not intersect, then they are parallel.

STEP 2   Create a truth table for the inverse.

p	q	$\sim p$	$\sim q$	$\sim p \rightarrow \sim q$
T	T	F	F	T
T	F	F	T	T
F	T	T	F	F
F	F	T	T	T

SOLUTION   **If two lines do not intersect, then they must be parallel. The truth values are shown above.**

COACHED EXAMPLE

Write the contrapositive of the statement "All frogs are amphibians."

THINKING IT THROUGH

To write the contrapositive, first write the statement in "if...then" form.

If an animal is a _____ then it is a(n) _____.

The contrapositive both _____ and _____ the hypothesis and conclusion.

In the conditional statement, the hypothesis is: _____.

The conclusion is: _____.

Then the contrapositive of the conditional is _____.

Lesson Practice

Choose the correct answer.

1. What is the converse of the statement "If it is 3:00 PM in New York, then it is 12:00 PM in California"?

    (1) If it is not 3:00 PM in New York, then it is not 12:00 PM in California.

    (2) If it is 12:00 PM in California, then it is 3:00 PM in New York.

    (3) If it is not 12:00 PM in California, then it is not 3:00 PM in New York.

    (4) If it is not 3:00 PM in New York, then it is 12:00 PM in California.

2. Let p be the hypothesis and q be the conclusion of a conditional statement. Which of the following correctly symbolizes the inverse of the conditional?

    (1) $\sim p \rightarrow \sim q$

    (2) $\sim q \rightarrow \sim p$

    (3) $q \rightarrow p$

    (4) $p \rightarrow q$

3. Given the conditional "If Gina practices the piano, then she will do well in the recital", which of the following is the inverse of the statement?

    (1) If Gina does not practice the piano, then she will not do well in the recital.

    (2) If Gina does not practice the piano, then she will do well in the recital.

    (3) If Gina does not do well in the recital, then she did not practice the piano.

    (4) If Gina does well in the recital, then she practiced the piano.

4. Given the following two statements, the second statement is what form of the first statement?

    First statement: If an angle has a measure less than 90°, then it is acute.

    Second statement: If an angle is not acute, then it does not have a measure less than 90°.

    (1) contrapositive

    (2) converse

    (3) inverse

    (4) negation

5. Given the statement "If the car will not run, then the battery is dead", what is the contrapositive?

    (1) If the battery is dead, then the car will not run.

    (2) If the car runs, then the battery is not dead.

    (3) If the car will not run, then the battery is not dead.

    (4) If the battery is not dead, then the car will run.

6. Given the following two statements, the second statement is what form of the first statement?

First statement: If the sky is cloudy, then it will rain.

Second statement: If it rains, then the sky is cloudy.

    (1) contrapositive

    (2) converse

    (3) inverse

    (4) negation

OPEN-ENDED QUESTION

7. Given the definition: An obtuse triangle has one obtuse angle.

    A. Write the definition as a conditional statement.

    B. Write the contrapositive of the conditional statement.

    C. Determine the truth value of the contrapositive.

Lesson 4

The Laws of Logic

G.G.27

The **Laws of Logic** are patterns of logical reasoning that use conditional statements.

Law of Detachment	
Explanation	If $p \rightarrow q$ is a true conditional statement and p is true, then q is true.
Symbolically	$p \rightarrow q$ p Conclusion: q
Example	• True conditional: If it is raining, then the ground is wet. • It is raining. • Conclusion (by the Law of Detachment): The ground is wet.

Law of Syllogism	
Explanation	If $p \rightarrow q$ and $q \rightarrow r$ are true conditional statements, then $p \rightarrow r$ is also a true conditional statement.
Symbolically	$p \rightarrow q$ $q \rightarrow r$ Conclusion: $p \rightarrow r$
Example	• True conditional: If it is raining, then the ground is wet. • True conditional: If the ground is wet, then the game will be canceled. • Conclusion (by the Law of Syllogism): If it is raining, then the game will be canceled.

EXAMPLE 1

Grace goes to New York City on a school field trip. If the following statements are true, can you conclude that Grace visits the Statue of Liberty?

- If Grace goes to New York City, then she will go to Liberty Island.

- If Grace goes to Liberty Island, then she will visit the Statue of Liberty.

STRATEGY   **Use the Laws of Logic.**

STEP 1   Let p, q, and r represent each statement.

p: Grace goes to New York City.

q: Grace goes to Liberty Island.

r: Grace visits the Statue of Liberty.

STEP 2   Use the Law of Syllogism.

$p \rightarrow q$ is true and $q \rightarrow r$ is true. Then (by the Law of Syllogism), $p \rightarrow r$ is true.

If Grace goes to New York City, then she will visit the Statue of Liberty.

STEP 3   Use the Law of Detachment.

The original problem tells us Grace goes to New York City. Therefore, p is true.

Step 2 concludes that $p \rightarrow r$ is also true. Then (by the Law of Detachment), r is also true. Therefore, Grace visits the Statue of Liberty.

SOLUTION   **The Law of Syllogism and Law of Detachment both allow you to conclude that Grace visits the Statue of Liberty.**

Law of Contrapositive		
Explanation	$p \rightarrow q$ is a true conditional statement if and only if $\sim q \rightarrow \sim p$ is a true conditional statement.	
Symbolically	$p \rightarrow q$ Conclusion: $\sim q \rightarrow \sim p$	$\sim q \rightarrow \sim p$ Conclusion: $p \rightarrow q$
Example	• True conditional: If it is raining, then the ground is wet. • Conclusion (by the Law of Contrapositive): If the ground is not wet, then it is not raining. and • True conditional: If the ground is not wet, then it is not raining. • Conclusion (by the Law of Contrapositive): If it is raining, then the ground is wet.	

Law of Modus Tollens (equivalent to using the Law of Contrapositive)	
Explanation	If $p \rightarrow q$ is a true conditional statement and $\sim q$ is true, then $\sim p$ is true.
Symbolically	$p \rightarrow q$ $\sim q$ Conclusion: $\sim p$
Example	• True conditional: If it is raining, then the ground is wet. • The ground is not wet. • Conclusion (by the Law of Modus Tollens): It is not raining.

Law of Double Negation	
Explanation	The Law of Negation states that the original statement and the negation of the negation of the original statement are the same thing.
Symbolically	$p \leftrightarrow \sim(\sim p)$
Example	• Original statement: It is raining. • Double Negation: It is not true that it is not raining. • Conclusion: It is raining.

EXAMPLE 2

Given the following, draw a valid conclusion and state the law of logic being applied.

- The lines do not intersect to form right angles.

- If two lines are perpendicular, then they intersect to form right angles.

STRATEGY   **Use p and q representation.**

STEP 1   Let p and q represent each statement in the conditional: If two lines are perpendicular, then they intersect to form right angles.

  p: Two lines are perpendicular.

  q: They intersect to form right angles.

STEP 2   Represent the statement "The lines do not intersect to form right angles" as $\sim q$.

STEP 3   Organize the given information. The following is the given information in the problem.

  $p \rightarrow q$

  $\sim q$

STEP 4   Look at the Laws of Logic that have been introduced to you. From the given information in the problem, you can conclude $\sim p$ by the Law of Modus Tollens.

SOLUTION   **By the Law of Modus Tollens, you can conclude $\sim p$:**
The two lines are not perpendicular.

DeMorgan's Laws	
Explanation	DeMorgan's Laws are rules of logic that show what happens when you negate a conjunction or negate a disjunction.
Symbolically	$\sim(p \wedge q) \leftrightarrow \sim p \vee \sim q$ and $\sim(p \vee q) \leftrightarrow \sim p \wedge \sim q$
Example	• It is not true that it is raining and the ground is wet. • Conclusion (by DeMorgan's Laws): It is not raining or the ground is not wet. • It is not true that either it is raining or the ground is wet. • Conclusion (by DeMorgan's Laws): It is not true that it is not raining and the ground is not wet.

Law of Disjunctive Interference		
Explanation	The Law of Disjunctive Interference states that if a disjunction is true and one of the statements is false, then the other statement must be true.	
Symbolically	$p \vee q$ $\sim p$ • Conclusion: q p • Conclusion: $\sim q$	$p \vee q$ $\sim q$ • Conclusion: p q • Conclusion: $\sim p$
Example	• True disjunction: It is raining or the neighbor is watering the lawn. • The neighbor is not watering the lawn. • Conclusion: It is raining.	

EXAMPLE 3

From the given information, draw a valid conclusion and state the law of reasoning being applied. One of the statements is false.

$p \vee q$

p

STRATEGY **Use the Laws of Logic.**

Look at the Laws of Logic that involve disjunctions. These are DeMorgan's Laws and the Law of Disjunctive Interference.

This given information can be found in the Law of Disjunctive Interference, because this law deals with the parts of a disjunction, not the whole disjunction.

SOLUTION **By the Law of Disjunctive Interference, given $p \vee q$ and p true you can conclude $\sim q$ because one of the statements is false.**

COACHED EXAMPLE

Assume the following statement is true and draw a logical conclusion. State the Law of Logic behind your reasoning.

It is not true that the solution is an acid and it is a base.

THINKING IT THROUGH

This statement is an example of a _____ because it uses the word_____.

Use p and q representation for the statement.

Let p be the statement: _____.

Let q be the statement: _____.

Then the complete given statement can be represented: $\sim$_____.

_____ are the laws of logic that show what happen when you negate a conjunction or disjunction.

One of these laws tells you that $\sim(p \wedge q) \leftrightarrow$ _____.

When you change this from p and q representation into wording, you can draw the conclusion:
_____.

Lesson Practice

Choose the correct answer.

1. For the following premise, draw a valid conclusion.

    $\sim(\sim p)$

    (1) p
    (2) q
    (3) $p \wedge q$
    (4) $\sim p$

2. Which is a valid conclusion?

    $\sim p \vee \sim q$

    (1) $\sim p \vee \sim q$
    (2) $\sim p \wedge \sim q$
    (3) $\sim(p \wedge q)$
    (4) $\sim(p \vee q)$

3. Given the following premises, what is the logical conclusion? Which Law of Logic is being applied?

    p

    $p \rightarrow q$

    (1) $\sim q$; Law of Contrapositive
    (2) $\sim p$: Law of Modus Tollens
    (3) q: Law of Syllogism
    (4) q: Law of Detachment

4. Which Law of Logic is being applied to draw the conclusion?

    Angela will be a veterinarian, or she will be a pediatrician.

    Angela will not be a pediatrician.

    Conclusion: Angela will be a veterinarian.

    (1) Law of Contrapositive
    (2) Law of Modus Tollens
    (3) DeMorgan's Laws
    (4) Law of Disjunctive Interference

5. Given the following statements, draw a valid conclusion:

    If the snowstorm hits Philadelphia, then it will hit New York City.

    If the snowstorm hits New York City, then it will hit Albany.

    (1) If the storm does not hit Philadelphia, then it will hit Albany.
    (2) If the storm hits Philadelphia, then it will hit Albany.
    (3) If the storm hits Albany, then it will not hit New York City.
    (4) If the storm does not hit New York City, then it will hit Albany.

OPEN-ENDED QUESTION

6. Given the statement: It is not true that Abraham Lincoln was the sixteenth president and he was the nineteenth president.

    A. Write a statement using p and q representation.

    _____

    B. Draw a logical conclusion from the given statement.

    _____

    C. What Laws of Logic did you apply to draw this conclusion?

    _____

5 Logic Proofs: Hypothesis to Conclusion

 G.G.27

In geometry, a **proof** is a logical argument to prove something true. A proof includes given information, or the hypothesis, and a statement of what is to be proved, or the conclusion. There may be many different stategies you can use to prove the same conclusion from the same given information.

Two-column proofs have a logical series of statements arguing to the conclusion. Each statement will be supported with a reason, which may include definitions or previously stated facts.

Direct proofs use deductive reasoning to build a true statement from other statements that have already been proven true.

EXAMPLE 1

Create a direct proof for the following.

Given: Two angles of a triangle have measures 35° and 55°.

Prove: The triangle is a right triangle.

STRATEGY   **Use a two-column proof.**

STEP 1   Create a table. Label the first column "statements" and the other column "reasons."

STEP 2   Begin the proof with the given information.

Statements	Reasons
Two angles of a triangle have measures 35° and 55°.	Given

STEP 3   Continue making logical arguments until you reach the conclusion.

Statements	Reasons
Two angles of a triangle have measures 35° and 55°.	Given
The third angle of the triangle has a measure of 90°.	The sum of the angles of a triangle is 180°.
The triangle is a right triangle.	Definition of a right triangle: A right triangle has exactly one right angle.

SOLUTION   **The complete proof is shown in Step 3.**

An **indirect proof** uses negation to go from hypothesis to conclusion. When creating an indirect proof, you assume the opposite of what you want to prove is true. Then use logical reasoning until you reach a contradiction of this statement. Since the assumption you made was false, then the original statement must be true. A proof does not always have to be in table format as in Example 1. A proof can also be a convincing argument.

EXAMPLE 2

Using indirect reasoning, prove that every quadrilateral must contain at least one acute or right angle.

STRATEGY   **Create an indirect proof.**

STEP 1   Write the negation of what you want to prove.

A quadrilateral does not contain any acute or right angles.

STEP 2   Make the next logical conclusion from this statement.

If a quadrilateral does not have any acute or right angles, then every angle must have a measure greater than 90°.

STEP 3   Make the next logical conclusion.

If every angle has a measure greater than 90°, then the sum of the angles of the quadrilateral is greater than 360°.

STEP 4   Look for a contradiction.

It is a fact that the sum of the angles of a quadrilateral must be 360°. Therefore, the original statement is false.

SOLUTION   **The negation is false, therefore, the original statement is true: Every quadrilateral must have at least one acute or right angle.**

A **tautology** is a statement that is always true. If you create a truth table for a tautology, the last column will be all true.

EXAMPLE 3

Use a truth table to prove that $p \rightarrow (q \rightarrow p)$ is a tautology.

STRATEGY   **Create a truth table.**

STEP 1   Complete the first two columns.

The first part of the truth table will have the truth values for p and q.

p	q
T	T
T	F
F	T
F	F

STEP 2   Complete the third column.

Using the "Order of Operations" the third column will contain $q \rightarrow p$ from inside the parenthesis. Recall from previous lessons how to find the truth values for this conditional. Switch the order of the p and q column to make it easier to solve for $q \rightarrow p$.

q	p	q → p
T	T	T
F	T	T
T	F	F
F	F	T

STEP 3   Complete the fourth column.

The final column of the truth table will have the truth values for $p \rightarrow (q \rightarrow p)$. Look only at the truth values between the first column (p) and the third column ($q \rightarrow p$).

q	p	q → p	p → (q → p)
T	T	T	T
F	T	T	T
T	F	F	T
F	F	T	T

SOLUTION   **As shown by all T's in the final column of the truth table, this conditional is always true; therefore, it is a tautology.**

COACHED EXAMPLE

One of DeMorgan's Laws tells us that $\sim(p \wedge q) \leftrightarrow \sim p \vee \sim q$. This means that $\sim(p \wedge q)$ and $\sim p \vee \sim q$ are logically equivalent; therefore, they have the same truth values.

Complete a truth table to prove $\sim(p \wedge q) \leftrightarrow \sim p \vee \sim q$.

THINKING IT THROUGH

First, create a truth table for $\sim(p \wedge q)$.

p	q	$p \wedge q$	$\sim(p \wedge q)$
T	T	_____	_____
T	F	_____	_____
F	T	_____	_____
F	F	_____	_____

Next, create a truth table for $\sim p \vee \sim q$.

p	q	$\sim p$	$\sim q$	$\sim p \vee \sim q$
T	T	_____	F	_____
T	F	_____	T	_____
F	T	_____	F	_____
F	F	_____	T	_____

Now compare the final columns of the two truth tables.

Since the two columns have the same _____, you have proven $\sim(p \wedge q) \leftrightarrow \sim p \vee \sim q$.

Lesson Practice

Choose the correct answer.

1. What is the first step of an indirect proof of the statement: ∠*a* is an acute angle?

    (1) ∠*a* is an acute angle.

    (2) ∠*a* is not an acute angle.

    (3) The measure of ∠*a* is less than 90°.

    (4) The measure of ∠*a* is not less than 90°.

2. Rajesh knows the library closes at 8:00 each evening. When he walks up to the doors of the library one evening, they are locked. Rajesh thinks, "If the time were before 8:00, the library would still be open. Because the library is closed, it must be after 8:00." What type of proof did Rajesh create with his reasoning?

    (1) detachment proof

    (2) direct proof

    (3) indirect proof

    (4) negation proof

3. Ti wanted to prove that *AC* = *AB* + *BC*. Which of the following is *not* a method he could use?

    (1) direct proof

    (2) indirect proof

    (3) two-column proof

    (4) tautology

4. Which direct proof correctly proves the following:

    Hypothesis: There is ice on the driveway.

    Conclusion: The temperature must be 32°F or less.

    (1) Because the freezing point of water is 32°F and there is ice on the driveway, the temperature outside must be 32°F or less.

    (2)

Statements	Reasons
The temperature is higher than 32°F.	Negation of conclusion
The freezing point of water is 32°F.	Previously proven fact
There is no ice on the driveway.	The Law of Modus Tollens

    (3)

Statements	Reasons
There is ice on the driveway.	Given
The freezing point of water is 32°F.	Previously proven fact
The temperature outside must be 32°F or less.	The Law of Detachment

    (4)

Statements	Reasons
There is ice on the driveway.	Given
The temperature outside must be 32°F or less.	Conclusion

5. Which of the following truth tables proves the Law of Double Negation?

(1)

p	q	p → q
T	T	T
T	F	F
F	T	T
F	F	T

(2)

p	~p	~(~p)
T	F	T
T	F	T
F	T	F
F	T	F

(3)

p	~p
T	F
T	F
F	T
F	T

(4)

p	q	p ∨ q	~(p ∨ q)
T	T	T	F
T	F	T	F
F	T	T	F
F	F	F	T

6. Identify the pair of statements that form a contradiction.

A. $\overline{LM} \cong \overline{OP}$

B. $\overline{LM} \parallel \overline{OP}$

C. $\overline{LM} \perp \overline{OP}$

D. The measure of $\overline{LM}$ is equal to the measure of $\overline{OP}$.

(1) A and B

(2) B and C

(3) C and D

(4) A and D

OPEN-ENDED QUESTION

7. Given: If two lines intersect, then they have a point in common.

    Conclusion: If two lines do not have any points in common, then they do not intersect.

    A. What Laws of Logic does this example demonstrate?

    _____

    B. Use truth tables to prove this law.

6 Inductive and Deductive Reasoning

G.G. 27

Deductive reasoning is the process of arguing from hypothesis to conclusion using given facts.

EXAMPLE 1

Many people use deductive reasoning as a regular part of their jobs. For example, a doctor confirms skin cancer through a biopsy, where the skin tissue is examined under a microscope to determine if cancer cells are present. Suppose a biopsy does not reveal cancer cells. Use deductive reasoning to make the conclusion the doctor may make.

STRATEGY   Use deductive reasoning.

It is a given fact that if the biopsy shows cancer cells, then the patient has skin cancer. If the biopsy does not show cancer cells, then the opposite conclusion can be drawn: the patient does not have skin cancer.

SOLUTION   **Through deductive reasoning, you can conclude the patient does not have skin cancer.**

Note: In actual medical practice, false negatives and false positives do occur and so the doctor might use other tests to verify his conclusion.

Another type of reasoning, **inductive reasoning**, draws conclusions from previous examples or patterns.

EXAMPLE 2

Vi observes birds at a bird feeder for an hour. Each goldfinch hangs upside down to eat. What conclusion might Vi make about the eating habits of goldfinches from this hour of observation?

STRATEGY   **Use inductive reasoning.**

Because inductive reasoning draws a conclusion from observed patterns, try to see what pattern there is among the goldfinches at the birdfeeder. Each goldfinch eats upside down.

SOLUTION   **Through inductive reasoning, Vi could conclude that goldfinches have the ability to eat hanging upside down.**

A **counterexample** to a statement is an example where the statement is not true. The hypothesis of the statement has been fulfilled, but the conclusion is false.

EXAMPLE 3

Prove this statement to be false: Alaska is the only state in the United States that begins and ends with the same letter.

STRATEGY   **Find a counterexample.**

You just need one counterexample to prove the statement is false. Find a state in the United States, other than Alaska, that begins and ends with the same letter.

Ohio is a state that begins and ends with the same letter.

SOLUTION   **The statement that Alaska is the only state in the United States that begins and ends with the same letter, is false because Ohio also begins and ends with the same letter.**

COACHED EXAMPLE

Identify each of the following as either deductive reasoning, inductive reasoning, or a counterexample.

A. The sum of the measures of two angles is 180°. You conclude the angles are supplementary.

B. An electronic store advertises, "You can't beat our prices!" You find the same television advertised for a lower price at a competitor's store.

C. Over the past ten years, an area has been hit with eight tornadoes. You conclude this is a likely spot for tornadoes to develop.

THINKING IT THROUGH

A. Look at the first example: The sum of the measures of two angles is 180°. You conclude the angles are supplementary.

It is a _____ that if the sum of two angles is ____, then the angles are _____.

_____ is a process of arguing from given facts to a _____.

Therefore, this is an example of _____.

B. Look at the next example: An electronics store advertises, "You can't beat our prices!" You find the same television advertised for a lower price at a competitor's store.

Because you found a lower price, you have proven the electronics store's claim to be _____.
A _____ is an example where a statement is not _____.

Therefore, this is a _____.

C. Look at the last example: Over the past ten years, an area has been hit with eight tornadoes. You conclude this is a likely spot for tornadoes to develop.

In this example, you found a _____ and used it to draw a _____.

_____ draws conclusions from _____.

Therefore, this is an example of _____.

Lesson Practice

Choose the correct answer.

1. Which of the following is an example of inductive reasoning?

   (1) After measuring each angle of a triangle and finding their measurements to be equal, you conclude it is an equilateral triangle.

   (2) After your car stalls, you look at the fuel gauge and determine you have run out of gas.

   (3) After twenty customers exit a building carrying food containers, you conclude that the building is a restaurant.

   (4) After taking the measurements of a driveway, a construction worker concludes how much concrete will be needed for the job.

2. Find a counterexample to this statement: Every city has, at most, one professional football team.

   (1) Albany does not have a professional football team.

   (2) New York City has two professional football teams.

   (3) Denver has one professional football team.

   (4) Some cities do not have professional football teams.

3. The following statement is an example of what type of reasoning?

   A resident of San Diego claims, "It never snows in San Diego." You check the almanac and find that it snowed in San Diego in 1967.

   (1) a counterexample

   (2) a contrapositive

   (3) deductive reasoning

   (4) inductive reasoning

4. Inductive and deductive reasoning are both ways to argue from what to a conclusion.

   (1) contrapositive

   (2) counterexample

   (3) hypothetical

   (4) hypothesis

5. Which of the following statements about counterexamples is false?

   (1) If you are able to find a counterexample to a statement, this proves the statement to be false.

   (2) A counterexample is one example where a statement is false.

   (3) In a counterexample, the hypothesis has been fulfilled, but the conclusion is false.

   (4) Several counterexamples are necessary to prove a statement false.

OPEN-ENDED QUESTION

6. When asked to define a square, Naomi said, "A square is a geometric figure with four right angles."

   A. Is this an acceptable definition?

   _____

   B. Find a counterexample to Naomi's definition.

   _____

7 Basic Postulates

A **postulate** is a statement that is accepted as true without proof. The following table contains some of the postulates used in proofs.

Postulate	Description	Example
Addition Postulate	If equal quantities are added to equal quantities, the sums are equal.	If $a = b$ and $c = d$, then $a + c = b + d$.
Subtraction Postulate	If equal quantities are subtracted from equal quantities, the differences are equal.	If $a = b$ and $c = d$, then $a - c = b - d$.
Multiplication Postulate	If equal quantities are multiplied by equal quantities, the products are equal.	If $a = b$ and $c = d$, then $a \cdot c = b \cdot d$.
Division Postulate	If equal quantities are divided by equal quantities, the quotients are equal.	If $a = b$ and $c = d$, then $a \div c = b \div d$, where c and d are not equal to 0.
Transitive Postulate	Quantities equal to the same or equal quantities are equal to each other. This also works for congruence.	If $a = b$ and $c = b$, then $a = c$. If $a \cong b$ and $c \cong b$, then $a \cong c$.
Substitution Postulate	A quantity may be substituted for its equal in any expression.	If $a = b$ and $b + c = d$, then $a + c = d$.
Symmetric Postulate	If one quantity is equal or congruent to a second quantity, then the second quantity is equal or congruent to the first quantity.	If $a = b$, then $b = a$. If $a \cong b$, then $b \cong a$.
Reflexive Postulate	Any quantity equals itself, or any quantity is congruent to itself.	$a = a$ $a \cong a$
Partition Postulate	The whole is equal to the sum of its parts.	$\overline{AB} + \overline{BC} = \overline{AC}$
Powers Postulate	Like powers of equal quantities are equal.	If $a = b$, then $a^n = b^n$.
Roots Postulate	Like roots of equal quantities are equal.	If $a = b$, then $\sqrt[n]{a} = \sqrt[n]{b}$.

EXAMPLE 1

Solve the equation $5x + y + 30 = 170$, given that $y = 2x$.
With each step, identify the postulate you used.

STRATEGY   **Solve the equation.**

STEP 1   Create a two-column table.

Complete the steps for solving the equation in the first column. In the second column, identify the postulate.

STEP 2   Complete the table.

Statements	Reasons
1. $5x + y + 30 = 170$	1. Given
2. $5x + (2x) + 30 = 170$	2. Substitution Postulate
3. $7x + 30 = 170$	3. Addition Postulate
4. $7x = 140$	4. Subtraction Postulate
5. $x = 20$	5. Division Postulate

SOLUTION   **The solution to the equation and the postulates used can be found in the table above.**

EXAMPLE 2

Identify the postulate that justifies each of the following statements.

(1) If $5a - 24 = 108$, then $5a = 132$.

(2) If $m\angle B = y$ and $y = 87°$, then $m\angle B = 87°$.

(3) If $BC = LM$, then $LM = BC$.

(4) If $\angle D \cong \angle E$ and $\angle F \cong \angle E$, then $\angle D \cong \angle F$.

STRATEGY   **Use the postulates from the table.**

STEP 1   Look at the first statement: If $5a - 24 = 108$, then $5a = 132$.

24 was added to both sides to get to the conclusion.

The Addition Postulate was used.

STEP 2   Look at the second statement: If $m\angle B = y$ and $y = 87°$, then $m\angle B = 87°$.

87° was substituted for y in the first expression.

The Substitution Postulate was used.

STEP 3   Look at the third statement: If $\overline{BC} \cong \overline{LM}$, then $\overline{LM} \cong \overline{BC}$.

The expressions were switched from one side of the equation to the other.

The Symmetric Postulate was used.

STEP 4   Look at the fourth statement: If $\angle D \cong \angle E$ and $\angle F \cong \angle E$, then $\angle D \cong \angle F$.

Both $\angle D$ and $\angle F$ are congruent to $\angle E$. Therefore, they are congruent to each other.

The Transitive Postulate was used.

SOLUTION   **(1) Addition Postulate**
(2) Substitution Postulate
(3) Symmetric Postulate
(4) Transitive Postulate

EXAMPLE 3

Given that $RS = TU$, the table proves that $RT = SU$. Complete the proof by stating the postulates used for each step.

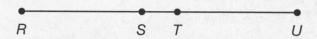

Statement	Reason
1. $RS = TU$	1. Given
2. $ST = ST$	2.
3. $RS + ST = ST + TU$	3.
4. $RS + ST = RT$	4.
5. $ST + TU = SU$	5.
6. $RT = SU$	6.

STRATEGY   **Complete the proof table.**

STEP 1   The first step of the proof is the given information. The second step states that a quantity equals itself: $ST = ST$. Check the list of postulates. This is the Reflexive Postulate.

STEP 2   In the third step, the same quantity, ST, is added to both sides of the equation: $RS + ST = ST + TU$. This is the Addition Postulate.

STEP 3   Look at the fourth and fifth steps of the proof: $RS + ST = RT$ and $ST + TU = SU$. For both steps, two parts are added to make the whole. This is the Partition Postulate.

STEP 4   Because $RS + ST = RT$ and $ST + TU = SU$, the quantities RT and SU are substituted for their equal quantities in the expression $RS + ST = ST + TU$.

SOLUTION   **The complete proof is shown below.**

Statement	Reason
1. $RS = TU$	1. Given
2. $ST = ST$	2. Identity Postulate
3. $RS + ST = ST + TU$	3. Addition Postulate
4. $RS + ST = RT$	4. Partition Postulate
5. $ST + TU = SU$	5. Partition Postulate
6. $RT = SU$	6. Substitution Postulate

COACHED EXAMPLE

Solve the equation $(x + y + z)^4 = (6x)^4$, given that $y = x + 3$ and $z = 2x$. With each step, identify the postulate you used. What are the values of x, y, and z?

THINKING IT THROUGH

Both sides of the equation are being raised to the same _____.

Since the like powers are equal, the quantities being raised to the powers are also equal.

This is the _____ Postulate.

This means $x + y + z =$ _____.

Substitute the values for y and z into this new equation: _____;
the _____ Postulate justifies this step.

Simplify by combining like terms: _____.

Subtract _____ from both sides of the equation to get the variable on one side.

This is the _____ Postulate.

The new equation is _____.

_____ both sides of the equation by _____ to isolate the variable.

This is the _____ Postulate.

You have solved the equation for x: $x =$ _____.

Put this value of x into the equations $y = x + 3$ and $z = 2x$ to find the values of y and z.

This is the _____ Postulate.

The solution for $(x + y + z)^4 = (6x)^4$, given that $y = x + 3$ and $z = 2x$ is
$y =$ _____ and $z =$ _____

Lesson Practice

Choose the correct answer.

1. Which postulate justifies the following statement?

    If $\angle W \cong \angle X$, then $\angle X \cong \angle W$.

    (1) Transitive Postulate

    (2) Symmetric Postulate

    (3) Substitution Postulate

    (4) Identity Postulate

2. Which postulate would you use in the first step of solving the equation $3a - 19 = 47$?

    (1) Addition Postulate

    (2) Division Postulate

    (3) Substitution Postulate

    (4) Subtraction Postulate

3. The following statement is an example of what postulate?

    If $\overline{AB} \cong \overline{CD}$ and $\overline{CD} \cong \overline{EF}$, then $\overline{AB} \cong \overline{EF}$.

    (1) Transitive Postulate

    (2) Symmetric Postulate

    (3) Powers Postulate

    (4) Partition Postulate

4. Any quantity is equal to itself. What postulate is this?

    (1) Transitive Postulate

    (2) Symmetric Postulate

    (3) Roots Postulate

    (4) Reflexive Postulate

5. The first two rows of a proof are shown below. What is the missing reason?

Statement	Reason
1. $m\angle 1 + m\angle 2 = 180$ and $180 = m\angle 3 + m\angle 4$	1. Given
2. $m\angle 1 + m\angle 2 = m\angle 3 + m\angle 4$	2. _____

    (1) Transitive Postulate

    (2) Symmetric Postulate

    (3) Roots Postulate

    (4) Identity Postulate

6. Which of the following statements about postulates is false?

    (1) The Addition Postulate, Subtraction Postulate, Multiplication Postulate, and Division Postulate are all examples of postulates.

    (2) A postulate has been previously proven true.

    (3) The Roots Postulate states like roots of equal quantities are equal.

    (4) A postulate is accepted as true without proof.

OPEN-ENDED QUESTION

7. At its top floor, the height of the Empire State Building is 1,250 feet. However, the pinnacle adds additional height to the building, as shown in the accompanying diagram. The height of the pinnacle is 47 feet less than one-fifth the height of the building from street level to the top floor.

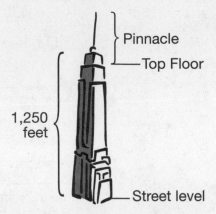

A. Let *p* represent the height of the pinnacle. Write an equation to find the height of the pinnacle.

_____

B. Create a two-column table. In the first column, you will complete the steps to solve the equation you wrote above and find the height of the pinnacle. In the second column, state the postulate that justifies each step.

Geometric Definitions and Postulates

 G.G. 27

Definitions use known words to explain the meaning of a new word. Here are some definitions you will encounter throughout geometry. Some may be familiar, and some may be new to you.

Terms Relating to Points and Lines	Definition	Example
Line	A line is a series of points extending in two directions without end.	$\overleftrightarrow{XY}$ X ⟵———•———————————————•———⟶ Y
Line Segment	A segment contains two endpoints and all of the points on the line in between them.	$\overline{XY}$ X •———————————————• Y
Congruent Segments	Two segments are congruent if they have the same measure.	X •——•——• Y A •——•——• B These symbols indicate congruent segments.
Ray	A ray is a part of a line consisting of one endpoint and all of the points on one side of the endpoint.	$\overrightarrow{XY}$ X •———————————————•———⟶ Y
Collinear	Points, rays, and segments that lie on the same line are collinear.	Points X, Y, and Z are collinear. X Y Z ⟵——•——•——————•——⟶
Opposite Rays	Opposite rays are collinear rays with the same endpoint. The rays form a line.	$\overrightarrow{YX}$ and $\overrightarrow{YZ}$ are opposite rays. X Y Z ⟵——•——————•——————•——⟶

Coplanar	Points, rays, and segments that lie on the same plane are coplanar.	Points X, Y, and Z are coplanar.
Intersecting Lines	Two lines intersect if they have exactly one point in common.	$\overleftrightarrow{WX}$ intersects $\overleftrightarrow{YZ}$ at point V.
Perpendicular Lines	Two lines are perpendicular if they intersect to form right angles. Symbol: $\perp$	$x \perp y$
Parallel Lines	Two lines are parallel if they are coplanar and do not have any points in common; they do not intersect. Symbol: $\parallel$	$x \parallel y$ These symbols indicate parallel lines.
Midpoint of a Segment	The midpoint of a segment is the point that divides the segment into two congruent segments.	Point Y is the midpoint of $\overline{XZ}$. These symbols indicate congruent segments.

Terms Relating to Angles	Definition	Example
Congruent Angles	Two angles are congruent if they have the same measure.	
Acute Angle	An angle whose measure is less than 90°.	
Right Angle	An angle whose measure is equal to 90°.	This symbol indicates a right angle.
Obtuse Angle	An angle whose measure is greater than 90° but less than 180°.	138°
Straight Angle	An angle whose measure is equal to 180°.	X Y Z
Adjacent Angles	Two angles are adjacent if they share a common vertex and common side, but have no common interior points.	common side common vertex 2 1 ∠1 and ∠2 are adjacent angles.
Complementary Angles	Two angles are complementary if the sum of their measures is 90°.	T 60° U V W 30° X Y ∠TUV and ∠WXY are complementary angles.
Supplementary Angles	Two angles are supplementary if the sum of their angles is 180°.	V 130° T U W 50° X Y ∠TUV and ∠WXY are supplementary angles.

Vertical Angles	Two angles are vertical if their sides are formed by intersecting lines.	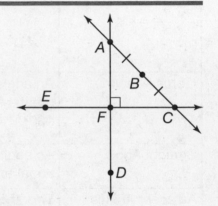
Linear Pair	Two angles are a linear pair if their noncommon sides are opposite rays that lie on the same line and are supplementary.	∠1 and ∠3, ∠2 and ∠4 on vertical angles. ∠1 and ∠2, ∠2 and ∠3, ∠3 and ∠4, ∠4 and ∠1 are all linear pairs.

EXAMPLE 1

Identify one of each of the following figures from the diagram.

   (a) perpendicular lines

   (b) vertical angles

   (c) acute angle

   (d) midpoint

   (e) adjacent angles

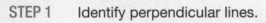

STRATEGY   **Use the definitions.**

  STEP 1   Identify perpendicular lines.

        The line that contains points A and D is perpendicular to the line that contains points E and C. $\overleftrightarrow{AD} \perp \overleftrightarrow{EC}$

  STEP 2   Identify vertical angles.

        There are several sets of vertical angles in the diagram. One set is ∠AFE and ∠CFD.

  STEP 3   Identify an acute angle.

        There are two angles which appear to have measures less than 90°. They are ∠FAB and ∠BCF.

  STEP 4   Identify a midpoint.

        $\overline{AB}$ and $\overline{BC}$ are marked with the same tick marks; this indicates that they are congruent. Therefore point B is a midpoint.

  STEP 5   Identify adjacent angles.

        There are several pairs of adjacent angles in the diagram. ∠EFA and ∠EFD share a side and therefore are adjacent.

SOLUTION   **The figures are identified in Steps 1–5.**

Many of the postulates you learned in the previous lesson relate mainly to algebra. The following table shows postulates that relate to geometry and geometric figures.

Postulates Relating to Lines and Segments	Description
Line	A line contains at least two points.
Plane	A plane contains at least three noncollinear points.
Unique Line	Through any two points, there is exactly one line.
Intersecting Lines	The intersection of two distinct lines is a point.
Unique Plane	Through any three noncollinear points, there is exactly one plane.
Intersecting Planes	The intersection of two distinct planes is a line.
Perpendicular Lines	Through any point not on a given line, exactly one perpendicular line can be drawn to the given line.
Parallel Lines	Through any point not on a given line, exactly one parallel line can be drawn to the given line.
Distance Postulate	For any two points on a line, and a given unit of measure, there is a unique positive number called the measure of the distance between the two points.
Midpoint Postulate	Every segment has exactly one midpoint.
Segment Addition Postulate	If three points A, B, and C are collinear, and B is between A and C, then $AB + BC = AC$.

EXAMPLE 2

Identify the postulate being illustrated in each of the diagrams below.

(1)

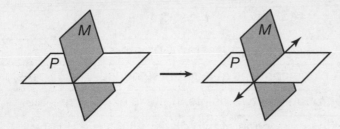

(2)

(3)

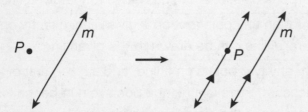

STRATEGY   **Compare the diagrams to the postulates.**

STEP 1   Look at the first set of diagrams.

The first part of the diagram illustrates plane *M* intersecting plane *P*. The second part of the diagram shows that the intersection is a line. This illustrates the intersecting planes postulate: The intersection of two distinct planes is a line.

STEP 2   Look at the next set of diagrams.

The first part of the diagram shows points *P* and *L*. The next diagram shows a line drawn through these two points. This illustrates the unique line postulate: Through any two points there is exactly one line.

STEP 3   Look at the last set of diagrams.

The first part of the diagram shows line *m* and point *P*, not on line *m*. The next set of diagrams shows a line drawn through *P* that is parallel to line *m*. This illustrates the parallel lines postulate: Through any point not on a given line, exactly one line can be drawn parallel to the given line.

SOLUTION   **(1) Intersecting Planes Postulate**
(2) Unique Line Postulate
(3) Parallel Lines Postulate

Postulates Relating to Angles	Description
Angle Addition Postulate	If *B* is in the interior of $\angle AOC$, then $m\angle AOB + m\angle BOC = m\angle AOC$.
Linear Pair Postulate	If two angles form a linear pair, then they are supplementary.

EXAMPLE 3

$\angle 1$ and $\angle 2$ form a linear pair. If $m\angle 1 = 32°$, find $m\angle 2$.

STRATEGY **Use the Linear Pair Postulate.**

STEP 1 If $\angle 1$ and $\angle 2$ are a linear pair, the Linear Pair Postulate states that they are supplementary.

Because the sum of supplementary angles is 180°, you can write the following equation:

$m\angle 1 + m\angle 2 = 180°$

STEP 2 Substitute any known values and solve.

$m\angle 1 + m\angle 2 = 180°$
$32° + m\angle 2 = 180°$
$m\angle 2 = 148°$

SOLUTION **The measure of $\angle 2$ is 148°.**

COACHED EXAMPLE

Use the rectangular prism below to answer the following questions.

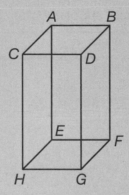

(1) Name 3 noncollinear points.

(2) What is the intersection of plane *BDGF* and plane *ACDB* ?

(3) Name the line segment containing point *B* that is perpendicular to $\overline{FG}$.

(4) Classify $\angle CHG$ as acute, right, or obtuse.

THINKING IT THROUGH

(1) Name 3 noncollinear points: If points are noncollinear, then they _____.
There are many sets of noncollinear points on this prism, but 3 are _____,
_____, and _____.

(2) Name the intersection of plane *BDGF* and plane *ACDB* by looking at the prism. Locate each plane
and find their intersection. The intersection of the two planes is _____.

(3) Name the line segment containing point *B* that is perpendicular to $\overline{FG}$:
The _____ postulate tells you that through any given point not on a line,
there is exactly one perpendicular to the given line.
Perpendicular means the lines _____ to form _____.

Locate $\overline{FG}$ on the prism. Now, locate point *B*. The line segment that contains *B* and is
perpendicular to $\overline{FG}$ is _____.

(4) Classify $\angle CHG$ as acute, right, or obtuse: Locate $\angle CHG$ on the prism.
It is the corner of one of the faces of the prism.
Since you are told that this is a rectangular prism, you know its faces are _____.
Each angle of a rectangle measures _____.
Therefore, m$\angle CHG =$ _____, and by definition, this angle is _____.

Lesson Practice

Choose the correct answer.

1. If M is between L and N on $\overleftrightarrow{LN}$, what ray is opposite $\overrightarrow{MN}$?

    (1) $\overrightarrow{NM}$
    (2) $\overrightarrow{ML}$
    (3) $\overrightarrow{LM}$
    (4) $\overrightarrow{LN}$

2. If point S is between points R and T on a line, which of the following demonstrates the Segment Addition Postulate?

    (1) $\overline{RS} + \overline{RT} = RT$
    (2) $\overline{RT} + \overline{RT} = RS$
    (3) $\overline{RS} + \overline{ST} = RT$
    (4) $\overline{ST} + \overline{RT} = RT$

3. If $m\angle 4 = 121°$ and $m\angle 5 = 59°$, then which of the following statements is not true?

    (1) $\angle 4$ and $\angle 5$ may form a linear pair.
    (2) $\angle 4$ and $\angle 5$ are supplementary.
    (3) $\angle 4$ and $\angle 5$ are complementary.
    (4) $\angle 4$ is obtuse.

4. Which of the following is not a postulate?

    (1) Every segment has exactly one midpoint.
    (2) Through any point not on a line, there is exactly one line parallel to the line.
    (3) For any two points on a line, there is a unique positive number called distance between the two points.
    (4) A plane contains at least two noncollinear points.

5. Which statement is not true concerning the following diagram?

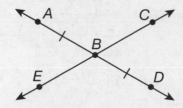

(1) Point B is the midpoint of $\overleftrightarrow{AD}$.

(2) $\angle ABE$ and $\angle ABC$ are vertical angles.

(3) $\angle CBD$ and $\angle DBE$ form a linear pair.

(4) By the angle addition postulate, $m\angle ABC + m\angle CBD = m\angle ABD$.

6. Let X be between W and Y on a line. Given the following information, use the segment addition postulate to solve for m.

$$WX = 2m + 2$$
$$XY = m + 3$$
$$WY = 4m$$

(1) $m = 3$

(2) $m = 4.5$

(3) $m = 5$

(4) $m = 6.5$

OPEN-ENDED QUESTION

7. Points $A(-5,3)$, $B(-2,0)$, and $C(4,0)$ are on a coordinate plane.

A. Classify $\angle ABC$.

_____

B. If point D is at $(0,3)$, write the angle addition postulate for $\angle ABC$.

_____

C. What is a coordinate for a point E that would create the straight angle $\angle EBC$?

_____

9 Perpendicular Lines and Planes

 G.G. 1, G.G. 2, G.G. 3, G.G. 4, G.G. 5, G.G. 6, G.G. 7

Recall that if lines or planes are perpendicular, they intersect to form right angles. Here are some additional postulates regarding perpendicular lines and planes.

Postulate	Definition	Example
Postulate 1	If a line is perpendicular to each of two intersecting lines at their point of intersection, then the line is perpendicular to the plane determined by them.	Line l is perpendicular to $\overleftrightarrow{AD}$ and $\overleftrightarrow{CB}$. Then it is also perpendicular to plane $ABCD$.
Postulate 2	Through a given point, there passes exactly one plane perpendicular to a given line.	Postulate 2: Given point P, only plane M is perpendicular to line l through this point.
Postulate 3	Through a given point, there passes exactly one line perpendicular to a given plane.	Postulate 3: Given point P, only line l is perpendicular to plane M through this point.
Postulate 4	Two lines perpendicular to the same plane are coplanar.	Lines l and n are perpendicular to plane M. Both lines l and n lie on plane O.

Postulate 5	Two planes are perpendicular to each other if and only if one plane contains a line perpendicular to the second plane.	Both line *l* and plane *O* are perpendicular to plane *M*.
Postulate 6	If a line is perpendicular to a plane, then any line perpendicular to the given line at its point of intersection with the given plane is on the given plane.	Line *l* is perpendicular to both line *n* and plane *M*.
Postulate 7	If a line is perpendicular to a plane, then every plane containing the line is perpendicular to the given plane.	Line *l*, plane *R*, and plane *O* are all perpendicular to plane *M*.

EXAMPLE 1

Which postulate is described by each of the following examples?

(a) An outdoor umbrella rests in the hole in the middle of the patio table.

(b) A worker is installing a picket fence. He has put two pickets into the ground at 90° angles to the ground.

(c) The axle of a wagon is perpendicular to the spokes of a wagon wheel.

STRATEGY   **Use the postulates in the table.**

STEP 1   Look at a: An outdoor umbrella rests in the hole in the middle of the patio table. Let the objects in the description represent points, lines, and planes.

The pole of the umbrella (line) and the table (plane) are perpendicular at the point where they meet, the hole in the table (point).

This describes Postulate 3: Through a given point (hole), there passes exactly one line (umbrella pole) perpendicular to a given plane (table).

STEP 2   Look at b: A worker is installing a picket fence. He has put two pickets into the ground at 90° angles to the ground. Again, let the objects in the description represent points, lines, or planes.

Each of the two pickets (lines) are perpendicular to the ground (plane). Additionally, both of the pickets are part of the fence (second plane).

This describes postulate 4: Two lines (pickets) perpendicular to the same plane (ground) are coplanar (both on the fence plane).

STEP 3   Look at c: The axle of a wagon is perpendicular to the spokes of a wagon wheel. Let the objects in the description represent points, lines, and planes.

The spokes (lines) of the wheel all intersect at the axle (line) and are perpendicular to the axle at this point. Let the wheel be the plane determined by all of the spokes.

This describes postulate 6: If a line (axle) is perpendicular to a plane (wheel), then any line (spokes) perpendicular to the given line at its point of intersection with the given plane (wheel) is on the given plane (wheel).

SOLUTION   **a) Postulate 3, b) Postulate 4, c) Postulate 6**

EXAMPLE 2

$\overleftrightarrow{ST}$ is perpendicular to plane M at point P. $\overleftrightarrow{ST}$ is also perpendicular to $\overleftrightarrow{WX}$ at point P. What can you conclude about $\overleftrightarrow{WX}$.

STRATEGY   **Use postulate 6.**

Postulate 6 relates to a line being perpendicular to both a plane and a line through the same point.

It states: If a line is perpendicular to a plane, then any line perpendicular to the given line at its point of intersection with the given plane is on the given plane.

Conclude that $\overleftrightarrow{WX}$ is on plane M.

SOLUTION $\overleftrightarrow{WX}$ **is on plane** M**.**

COACHED EXAMPLE

Use the diagram of a cube below to illustrate:

(1) Postulate 1

(2) Postulate 2

(3) Postulate 5

(4) Postulate 7

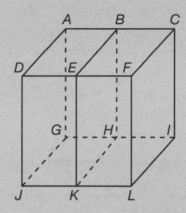

THINKING IT THROUGH

(1) Refer to Postulate 1:

Let the given line be $\overleftrightarrow{LI}$. Name two lines that $\overleftrightarrow{LI}$ is perpendicular to at their point of intersection:

_____ and _____.

Both of these lines are on plane _____.

Is $\overleftrightarrow{LI}$ perpendicular to this plane? _____

Then $\overleftrightarrow{LI}$ is perpendicular to lines _____ and _____ and to plane _____.

(2) Refer to Postulate 2:

Let the given point be *E* and the given line be $\overleftrightarrow{EK}$. Then, through *E*, the only plane perpendicular to $\overleftrightarrow{EK}$ is plane _____.

(3) Refer to Postulate 5:

Using two perpendicular planes, plane *ACFD* and plane *AGJD*, choose a line on *AGJD* that is also perpendicular to the other plane: _____.

Both the plane _____ and the line _____ that lies on this plane are perpendicular to plane _____.

(4) Refer to Postulate 7:

Let $\overleftrightarrow{FD}$ be the given line. This line is perpendicular to plane _____.

Now, name two planes that contain $\overleftrightarrow{FD}$: plane _____ and plane_____.

Are both of these planes perpendicular to this plane also? _____

Then $\overleftrightarrow{FD}$, plane_____, and plane _____ are all perpendicular to plane _____.

Lesson Practice

Choose the correct answer.

1. Which postulate is illustrated by the following diagram?

    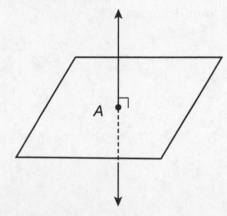

    (1) Postulate 3
    (2) Postulate 4
    (3) Postulate 5
    (4) Postulate 6

2. Two lines perpendicular to the same plane are

    (1) collinear.
    (2) coplanar.
    (3) opposite rays.
    (4) supplementary.

3. The sides of a door frame intersecting with the floor could be an example of which postulate?

    (1) Postulate 3
    (2) Postulate 4
    (3) Postulate 5
    (4) Postulate 6

4. Which of the following is not a postulate?

    (1) If a line is perpendicular to a plane, then every plane containing the line is perpendicular to the given plane.
    (2) Through a given point, there passes exactly one line perpendicular to the given plane.
    (3) Through a given point, there passes exactly one plane perpendicular to a given line.
    (4) Through a given point, there passes at least one line perpendicular to the given plane.

5. Complete this postulate: If a line is perpendicular to a plane, then every _____ containing the line is perpendicular to the given plane.

    (1) point
    (2) ray
    (3) segment
    (4) plane

6. Two planes are perpendicular if and only if one plane contains a _____ that is perpendicular to the second plane. Which of the following words best completes this postulate?

    (1) point
    (2) ray
    (3) line
    (4) plane

7. Given: $\overleftrightarrow{AB}$ lies on plane R, and $\overleftrightarrow{AB}$ is perpendicular to plane W.

   Which conclusion can you draw from the given information?

   (1) $\overleftrightarrow{AB}$ also lies on plane W.

   (2) Plane R and plane W are the same plane.

   (3) Plane R is perpendicular to plane W.

   (4) $\overleftrightarrow{AB}$ is not perpendicular to plane W.

8. Through point P, how many planes are perpendicular to $\overleftrightarrow{RS}$?

   (1) zero

   (2) one

   (3) two

   (4) infinitely many

OPEN-ENDED QUESTION

9. Given:

   $\overleftrightarrow{AB}$ and $\overleftrightarrow{CD}$ intersect at point P.

   $\overleftrightarrow{EF} \perp \overleftrightarrow{AB}$

   $\overleftrightarrow{EF} \perp \overleftrightarrow{CD}$

   $\overleftrightarrow{AB}$ and $\overleftrightarrow{CD}$ lie on plane M.

   A. Write the conclusion that can be drawn from the given information.

   _____

   _____

   _____

   B. What postulate did you use in A? Sketch a drawing to illustrate this postulate.

Lesson

10 Parallel Lines and Planes

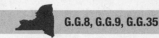
G.G.8, G.G.9, G.G.35

A **transversal** is a line that intersects two or more coplanar lines at different points. In the figure below, transversal *t* intersects lines *r* and *s*.

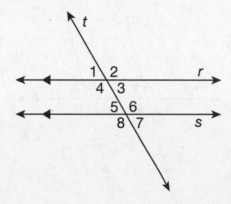

When a transversal intersects the two lines, several types of angles are created.

Corresponding angles, such as ∠2 and ∠6, occupy corresponding positions.

Alternate interior angles, such as ∠3 and ∠5, lie between *r* and *s* on opposite sides of the transversal.

Alternate exterior angles, such as ∠2 and ∠8, lie on the outside of *r* and *s* on opposite sides of the transversal.

Consecutive interior angles, such as ∠3 and ∠6, lie on the inside of *r* and *s* on the same side of the transversal.

Consecutive exterior angles, such as ∠2 and ∠7, lie on the outside of r and s on the same side of the transversal.

Also, recall from Lesson 8, vertical angles, such as ∠2 and ∠4, have sides that form two pairs of opposite rays.

EXAMPLE 1

Identify each of the following pairs of angles as corresponding, alternate interior, alternate exterior, consecutive interior, or vertical.

(a) ∠6 and ∠7

(b) ∠1 and ∠8

(c) ∠2 and ∠5

(d) ∠4 and ∠5

(e) ∠4 and ∠8

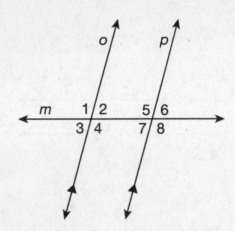

STRATEGY   **Use the definitions.**

STEP 1   Identify the line that is the transversal.

Line *m* is the transversal.

STEP 2   Identify ∠6 and ∠7.

Their sides form opposite rays. By definition, they are vertical angles.

STEP 3   Identify ∠1 and ∠8.

They are on the outside of lines *o* and *p*, but on opposite sides of the transversal. By definition, they are alternate exterior angles.

STEP 4   Identify ∠2 and ∠5.

They are on the inside of lines *o* and *p* and on the same side of the transversal. By definition, they are consecutive interior angles.

STEP 5   Identify ∠4 and ∠5.

They are on the inside of lines *o* and *p*, but on opposite sides of the transversal. By definition, they are alternate interior angles.

STEP 6   Identify ∠4 and ∠8.

They are in corresponding positions. By definition, they are corresponding angles.

SOLUTION   **(a) vertical angles (b) alternate exterior angles (c) consecutive interior angles (d) alternate interior angles (e) corresponding angles**

Two lines are parallel if they are coplanar and do not have any points in common; they do not intersect.

Unlike postulates, which are accepted without proof, **theorems** are conjectures that have been proven. The following table contains some postulates and theorems about given pairs of angles formed when two parallel lines are cut by a transversal.

Corresponding Angles Postulate: If two parallel lines are cut by a transversal, then the corresponding angles are congruent.

Alternate Interior Angles Theorem: If two parallel lines are cut by a transversal, then the alternate interior angles are congruent.

Consecutive Interior Angles Theorem: If two parallel lines are cut by a transversal, then the pairs of consecutive interior angles are supplementary.

Alternate Exterior Angles Theorem: If two parallel lines are cut by a transversal, then the alternate exterior angles are congruent.

Vertical Angles Theorem: If two angles are vertical, then they are congruent.

EXAMPLE 2

Complete this two-column proof. What theorem does this prove?

Given: $m \parallel n$
Prove: $\angle 2 \cong \angle 8$

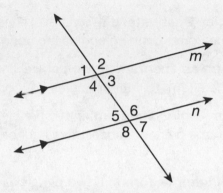

STRATEGY **Write a two-column proof.**

STEP 1 Determine the relationships between the angles.

From the Vertical Angles Theorem, you know that vertical angles are congruent. Make this the next step of the proof.

STEP 2 Identify the next step in the proof.

The Corresponding Angles Postulate tells you that if two lines are parallel, their corresponding angles are congruent. Use this for the next step of the proof.

STEP 3 Complete the proof.

$\angle 2 \cong \angle 4$ and $\angle 4 \cong \angle 8$. Then by the Transitive Postulate, $\angle 2 \cong \angle 8$.

Statements	Reasons
$m \perp n$	Given.
$\angle 2 \cong \angle 4$	Vertical angles are congruent.
$\angle 4 \cong \angle 8$	If two lines are parallel, corresponding angles are congruent.
$\angle 2 \cong \angle 8$	Transitive Postulate

STEP 4 Identify the theorem proved.

You have proved that $\angle 2$ is congruent to $\angle 8$, given that line m is parallel to line n. $\angle 2$ and $\angle 8$ are alternate exterior angles. Therefore, you have proved the Alternate Exterior Angles Theorem.

SOLUTION **The complete proof is shown in Step 3.**
This Example proved the Alternate Exterior Angles Theorem.

The following are two additional postulates about parallel lines and planes.

> **Postulate 8**: If a plane intersects two parallel planes, then the intersections are two parallel lines.
>
> **Postulate 9**: If two planes are perpendicular to the same line, then the planes are parallel.

EXAMPLE 3

Identify the postulate illustrated in each illustration.

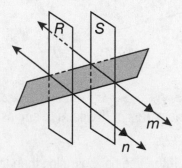

plane R ∥ plane S

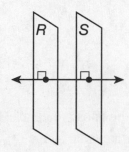

plane R ∥ plane S

STRATEGY   **Use the definitions.**

STEP 1   Identify the first illustration.

It shows two parallel planes being intersected by a third plane. It also shows that the lines of intersection, line m and line n, are parallel. Look at the postulates in the table above. This describes Postulate 8.

STEP 2   Identify the second illustration.

It shows two planes being intersected by a line to form right angles. It also shows the planes to be parallel. Look at the postulates in the table above. This describes Postulate 9.

SOLUTION   **Postulate 8 is illustrated in the first illustration; Postulate 9 is illustrated in the second illustration.**

COACHED EXAMPLE

If m∠1 = 97° and m∠7 = 83°, is line *s* parallel to line *t*?

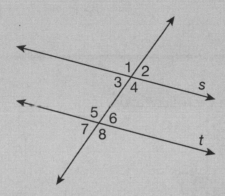

THINKING IT THROUGH

You can determine if the lines are parallel based on the measures of the pairs of angles created when line *s* and line *t* are intersected by the _____.

Based on the _____ Postulate, if the lines are parallel, then the corresponding angles are _____.

The corresponding angle to ∠1 is _____.

Find the measure of this angle.

∠5 and ∠7 are _____. This means, the sum of their measures is _____.

Since m∠7 = 83°, 83° + m∠5 = _____.

Solve this equation to find m∠5: _____.

Is m∠5 equal to m∠1? _____

Since their measures are equal, ∠5 and ∠1 are what type of angles? _____.

Since the corresponding angles are _____, line *s* and line *t* are _____.

Lesson Practice

Choose the correct answer.

1. Line l and line m are intersected by a transversal. The angles on the inside of l and m and on the same side of the transversal are

   (1) vertical angles.

   (2) corresponding angles.

   (3) consecutive interior angles.

   (4) alternate interior angles.

2. What is the value of x?

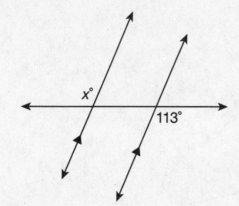

   (1) 23

   (2) 67

   (3) 113

   (4) The value of x cannot be determined by the given information.

3. Classify $\angle 1$ and $\angle 8$.

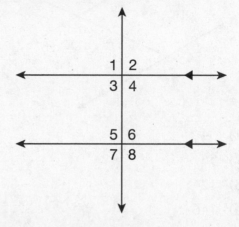

   (1) vertical angles

   (2) corresponding angles

   (3) consecutive angles

   (4) alternate exterior angles

4. Given: plane $Q \perp$ line l and plane $R \perp$ line l. What conclusion can you make from the given information?

   (1) plane $Q \perp$ plane R

   (2) plane $Q \parallel$ plane R

   (3) plane $Q \cong$ plane R

   (4) Line l lies on plane Q.

5. If a plane intersects two parallel planes, then the intersections are

   (1) a line.

   (2) two parallel lines.

   (3) one perpendicular line and one parallel line.

   (4) a plane.

6. Find the value of *a*.

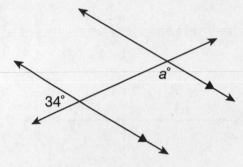

(1) 34

(2) 56

(3) 124

(4) 146

7. If two lines are intersected by a transversal so that the vertical angles are congruent, what can you conclude about the lines?

(1) The lines are parallel.

(2) The lines are perpendicular.

(3) The lines are on different planes.

(4) You cannot conclude anything about the lines because vertical angles are always congruent.

OPEN-ENDED QUESTION

8. Given: line *l* ‖ line *m*

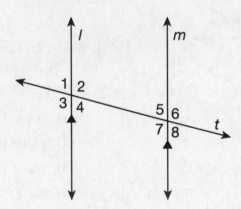

A. Prove: ∠5 ≅ ∠4.

_____

_____

_____

B. State the theorem that this proof demonstrates.

_____

Prisms

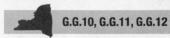
G.G.10, G.G.11, G.G.12

A **prism** is a polyhedron that has two parallel, congruent faces called **bases**. The other faces are **lateral faces**. A prism is named after the shape of its bases.

EXAMPLE 1

Name all sets of congruent faces of this rectangular prism.

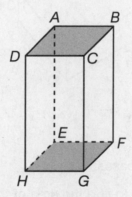

STRATEGY   Find congruent bases and lateral faces.

    STEP 1   Identify bases.

            In a prism, the bases are congruent.

            $ABCD \cong EFGH$

    STEP 2   Identify the congruent lateral faces.

            All the faces have the same height because they are intersected by parallel planes in addition, $CBFG$ and $ADHE$ have the same width so they are congruent. $BAEF$ and $DCGH$ also have the same width so they are congruent. Therefore, $CBFG \cong ADHE$ and $BAEF \cong DCGH$.

SOLUTION   *ABCD* is congruent to *EFGH*; *CBFG* is congruent to *ADHE*; and *BAEF* is congruent to *DCGH*.

The **altitude** of a prism is a perpendicular segment that connects the planes of the bases. The **height**, h, is the length of the altitude. The **volume of a prism** is the amount of cubic units contained within it. To calculate the volume, multiply the area of the base, B, by the height, h.

Volume of a Prism

$V = Bh$

EXAMPLE 2

A triangular prism is shown. What is the volume of the prism?

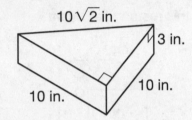

STRATEGY   Use the formula $V = Bh$.

   STEP 1   Identify the bases.

              The triangles are the bases.

   STEP 2   Use the formula $B = \frac{1}{2}bh$ to find the area of the base.

              Note that h is the height of the triangle, which is always perpendicular to the base of the triangle. The perpendicular sides, which each measure 10 inches, are the base and height of the triangle.

              $B = \frac{1}{2}bh = \frac{1}{2}(10)(10) = 50$

   STEP 3   Identify the altitude of the prism.

              The altitude of the prism is 3 inches.

   STEP 4   Substitute the values of the area of the base and the altitude into the volume formula.

              $V = Bh = (50)(3) = 150$

SOLUTION   **The volume of the triangular prism is 150 in³.**

EXAMPLE 3

Jessica cut out the cardboard net below to make a rectangular prism to use as a box for a ring. What is the volume of the box, rounded to the nearest tenth of an inch?

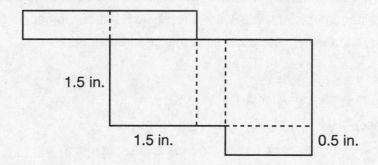

1.5 in.

1.5 in.   0.5 in.

STRATEGY Use the formula $V = Bh$.

STEP 1   Find the formula for the base area, B, and substitute into the volume formula.

Because the figure is a rectangular prism:

B = length · width = lw = (1.5 in)(1.5 in) = 2.25 in².

h = 0.5 in

STEP 2   Substitute the values into the formula and evaluate.

$V = Bh$

$V = lwh$

$V = (1.5)(1.5)(0.5)$

$V = 1.125$ in³.

SOLUTION The volume of the prism, rounded to the nearest tenth, is **1.1 in³**.

COACHED EXAMPLE

Marcus has two empty cube-shaped containers with sides 7 inches and 9 inches respectively. He fills the smaller container completely with water then pours it into the larger container. How deep, to the nearest tenth of an inch, will the water be in the larger container?

THINKING IT THROUGH

First calculate the volume of the smaller container.

Volume = (area of base) · (_____).

Because each container is a cube, each face of the cube is the shape of a _____. Therefore, the length, width, and height will all have the same measurement.

The area of the base of the smaller cube = side^2 = _____.

The height of the smaller cube is _____.

So the volume of the smaller cube is (_____)(_____) = _____.

When poured into the larger container, the water will have the same volume. You know that the water will cover the base of the larger container, but you need to find the height.

First find the area of the base of the larger cube.

Area of the base of the larger cube = side^2 = _____.

Because you now know the volume of the water and the area of the base, let h be the height of the water and make the substitutions into the formula:

Volume of the smaller cube = (base area of the larger cube) · (altitude of the water in the larger cube).

_____ = (_____)h

Solve for h and round to the nearest tenth.

The depth of the water in the larger container is about_____.

Lesson Practice

Choose the correct answer.

1. The length, width, and height of a rectangular prism are each tripled. What will happen to the volume of the prism?

    (1) The volume of the prism will be three times the original volume.

    (2) The volume of the prism will be nine times the original volume.

    (3) The volume of the prism will be eighteen times the original volume.

    (4) The volume of the prism will be twenty-seven times the original volume.

2. What is the volume of a cube with a side length of 8 cm?

    (1) 64 cm^3

    (2) 144 cm^3

    (3) 256 cm^3

    (4) 512 cm^3

3. Morgan fills a shoe box with sand. The area of the base of the shoe box is 48 in.^{2}, and the height is 6 in. She wants to transfer the sand to a mold of a triangular prism. If the area of the triangular base is 48 in.^{2}, what is the minimum height that will allow the prism to hold all of the sand?

    (1) 6 in.

    (2) 8 in.

    (3) 10 in.

    (4) 12 in.

4. There is 48 cm^3 of gelatin in the gelatin mold shown below. At what height is the gelatin?

    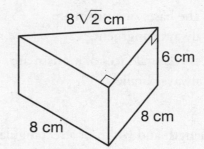

    (1) 0.75 cm

    (2) 1.5 cm

    (3) 2 cm

    (4) 2.25 cm

5. Which of the following is an expression for the volume of the prism in the figure?

    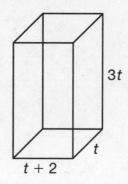

    (1) $4t + 2$

    (2) $3t^3 + 3t + 2$

    (3) $3t^3 + 6t^2$

    (4) $14t^2 + 16t$

6. Which of the following statements about prisms is false?

(1) A prism may have any number of sides.

(2) The bases of a prism are always parallel.

(3) The bases of a prism are always congruent.

(4) The lateral faces of a prism are always parallel.

7. The length and width of a rectangular prism are doubled and the height is quadrupled. The volume of the new prism will be how many times the volume of the original prism?

(1) 4

(2) 8

(3) 16

(4) 32

8. What is the volume of this rectangular oblique prism?

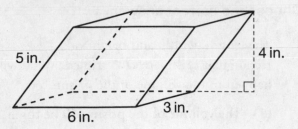

(1) 64 in.^{3}

(2) 72 in.^{3}

(3) 90 in.^{3}

(4) 124 in.^{3}

OPEN-ENDED QUESTION

9. Elijah built a box by cutting 2-inch squares from the corners of a rectangular sheet of cardboard, as shown in the diagram, and then folded the sides up. The volume of the box is 160 cubic inches, and the longer side of the box is 2 inches more than the shorter side.

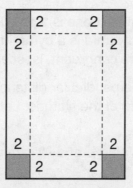

A. Let *x* represent the shorter side of the box. Write an equation to show the volume of the box.

_____

B. Find the number of inches in the shorter side of the cardboard before the corners were removed.

_____

_____

_____

Regular Pyramids

G.G 13

A **pyramid** is a polyhedron in which the base is a polygon, and the lateral faces are triangles with a **common vertex**. A regular pyramid is a pyramid whose base is a regular polygon. The lateral faces of a regular pyramid are congruent, isosceles triangles.

The altitude of the pyramid is the perpendicular distance between the base of the pyramid and the vertex. The height h is the length of the altitude. Let B represent the area of the base.

> **Volume of a Pyramid**
> $$V = \frac{1}{3}Bh$$

The diagram below shows that the volume of a pyramid is one-third the volume of a prism having the same base and the same height. $V = \frac{1}{3}Bh$

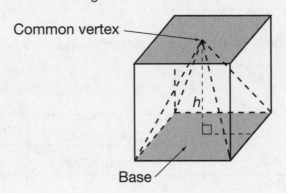

Common vertex

Base

EXAMPLE 1

What is the volume of the pyramid below?

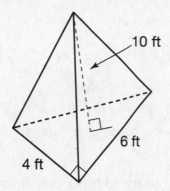

10 ft

6 ft

4 ft

STRATEGY   Use the formula $V = \frac{1}{3}Bh$.

     STEP 1   Find the area of the base of the prism.

                  The shape of the base is a right triangle.

                  Therefore, the area of the base is equal to $\frac{1}{2}bh$.

                  $B = \frac{1}{2}(6)(4) = 12 \text{ ft}^2$

     STEP 2   Substitute the values into the formula for the volume of a pyramid.

                  $V = \frac{1}{3}Bh$

                  $V = \frac{1}{3}(12)(10)$

                  $V = 40 \text{ ft}^3$

SOLUTION   The volume of the pyramid is 40 ft³.

EXAMPLE 2

The Louvre Pyramid is a modern, metal and glass pyramid built in 1989 in France. The square base of the Louvre Pyramid has sides 35 meters long, and it reaches 20.6 meters high. What is the volume of this pyramid to the nearest cubic meter?

STRATEGY   Use the formula $V = \frac{1}{3}Bh$.

STEP 1   Find the area of the base of the prism.

The shape of the base is a square.

Therefore, the area of the base is equal to s^2.

$B = (35)(35) = 1{,}225 \text{ m}^2$

STEP 2   Substitute the values into the formula for the volume of a pyramid.

$V = \frac{1}{3}Bh$

$V = \frac{1}{3}(1{,}225)(20.6)$

$V = 8412 \text{ m}^3$

SOLUTION   The volume of the Louvre Pyramid is 8,412 m^3 to the nearest cubic meter.

COACHED EXAMPLE

A pyramid has a hexagonal base with an area of 122 square centimeters. The volume of the pyramid is 488 cubic centimeters. What is the height of the pyramid?

THINKING IT THROUGH

The formula for the volume of a pyramid is _____.

From the given information, the area of the base is _____,
and the volume of the pyramid is _____.

Let h represent the unknown height and substitute the given information into the volume formula:
_____.

Finally, solve for h.
_____

The height of the hexagonal pyramid is _____.

Lesson Practice

Choose the correct answer.

1. Which statement about regular pyramids is true?

    (1) Regular pyramids have only triangle or square-shaped bases.

    (2) The lateral faces of a regular pyramid are congruent.

    (3) The volume of a regular pyramid is three times the area of the base.

    (4) The altitude of a regular pyramid is the parallel distance from the vertex to the center of the base.

2. A net for a square pyramid is pictured below. When the net is folded and taped to form a pyramid, calculate the volume of the solid.

    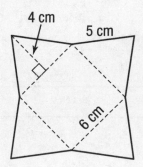

    (1) 48 cm^3

    (2) 60 cm^3

    (3) 120 cm^3

    (4) 144 cm^3

3. For a fund-raising event Jen filled a mold of a house with chocolate to use as an auction item. The mold is pictured below. When melted, the chocolate bars Jen used to fill the mold each had a volume of 8 cubic inches. How many chocolate bars did Jen have to melt to fill the mold?

    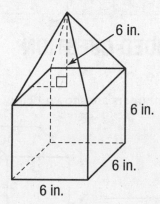

    (1) 10

    (2) 15

    (3) 20

    (4) 36

4. The base of a regular pyramid is an octagon with an area of 45 square inches. If the volume of the pyramid is 75 cubic inches, what is the height of the pyramid?

    (1) 1.67 inches

    (2) 5 inches

    (3) 10 inches

    (4) 30 inches

5. A regular square pyramid has an altitude of 15 centimeters, and a volume of 8,000 cubic centimeters. What is the length of one side of the square base, to the nearest centimeter?

(1) 23 centimeters

(2) 36 centimeters

(3) 40 centimeters

(4) 45 centimeters

6. A regular square pyramid has a base with side x and height $3x$. In terms of x, what is the volume of the pyramid?

(1) $3x^2$

(2) $3x^3$

(3) $3x^3 + x^2$

(4) x^3

OPEN-ENDED QUESTION

7. Two containers are shown below.

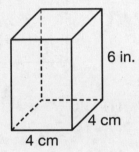
6 in.
4 cm
4 cm

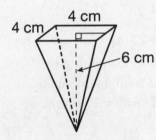
4 cm
4 cm
6 cm

A. What is the relationship between the volumes of the two containers?

_____

_____

B. Find the volume of the two containers.

_____

_____

_____

_____

_____

13 Cylinders

A **cylinder** has two congruent bases, like a prism, but the bases are circles. The altitude of a cylinder is the perpendicular distance between the bases. The height, h, is the length of the altitude. Also like a prism, the volume of a cylinder can be found by multiplying the area of the base, B, by the altitude.

> **Volume of a Cylinder**
> $$V = Bh = \pi r^2 h$$

EXAMPLE 1

Write an expression that represents the volume of the figure shown below.

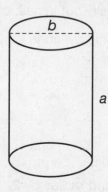

STRATEGY   **Identify the base and variables.**

STEP 1   Express the area of the base in terms of the variables shown in the figure.

The figure is a cylinder; therefore, the base is a circle.

The radius is half of the diameter. Since the diameter is b, $r = \frac{1}{2}b$.

$B = \pi r^2$ area of a circle

$B = \pi\left(\frac{1}{2}b\right)^2$

STEP 2   Identify the height.

The altitude is the perpendicular height of the cylinder.

height $= a$

STEP 3   Substitute the expressions for the area of the base and the altitude into the volume formula and simplify.

$V = Bh$

$V = \pi\left(\frac{1}{2}b\right)^2 a$

$V = \pi\frac{1}{4}b^2 a$

SOLUTION   **An expression that represents the volume of this cylinder is $\frac{1}{4}\pi b^2 a$.**

EXAMPLE 2

The height and radius of a cylinder are each cut in half. How does the volume of the prism change?

STRATEGY   **Pick sample measurements and compare.**

STEP 1   Pick numbers for the radius and the height.

Because 2 and 4 are easily cut in half, choose $h = 2$ and $r = 4$.

STEP 2   Find the volume of the original cylinder.

$V = Bh$

$V = \pi 4^2 2$

$V = \pi(16)(2)$

$V = \pi 32$ or 32π

STEP 3   Find the new height and radius.

Divide both the height and radius by 2.

$h = 1$ and $r = 2$

STEP 4   Find the volume of the new cylinder.

$V = Bh$

$V = \pi(2^2)1$

$V = \pi(4)(1)$

$V = \pi 4$ or 4π

STEP 5   Compare the original cylinder to the new cylinder.

The new volume, 4π, is one-eighth of the original volume 32π.

SOLUTION   **The new volume is $\frac{1}{8}$ the original volume.**

Imagine unrolling the lateral surface of a right cylinder; this surface forms a rectangle. The length of the rectangle is the circumference of the circular base of the cylinder. The width of the rectangle is the height of the cylinder. Therefore, the lateral area of a right cylinder equals the product of the height and the circumference of the base.

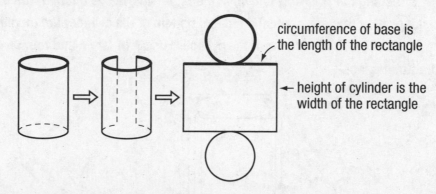

circumference of base is the length of the rectangle

height of cylinder is the width of the rectangle

Lateral Area of a Cylinder

$LA = 2\pi rh = \pi dh$, where r is the radius and d is the diameter

EXAMPLE 3

Clark wants to paint the exterior walls of the grain silo at his farm. The silo is in the shape of a cylinder and is 40 feet tall and 14 feet wide. Clark needs to calculate the square footage of the exterior of his silo in order to buy enough paint. To the nearest tenth, what is the area of the exterior of his silo?

STRATEGY   **Use the lateral surface area formula.**

    STEP 1   Use the formula $LA = \pi dh$. Identify the values of the variables.

            $d = 14$

            $h = 40$

    STEP 2   Substitute the diameter and height into the formula for the lateral area and solve.

            Let 3.14 be an approximation for π.

            $LA = (3.14)(14)(40) = 1{,}758.4$

SOLUTION   **The area of the exterior of the silo is approximately 1,758.4 ft².**

COACHED EXAMPLE

A cylinder is inside a rectangular prism, as shown in the figure below. The outside wall of the cylinder is touching the inside wall of the prism, and the top and bottom of the cylinder are touching the top and bottom of the prism. What is the volume of the cylinder? Use 3.14 for π, and round your answer to the nearest hundredth of a foot.

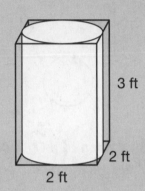

3 ft

2 ft

2 ft

THINKING IT THROUGH

Because the side of the cylinder is touching the side of the prism, the _____ of the cylinder is the same as the _____ of the prism, which is _____ ft.

Then the radius of the cylinder is _____ ft.

Because the top and bottom of the cylinder are touching the top and bottom of the prism, the _____ of the cylinder is the same as the _____ of the prism, which is _____ ft.

To find the volume of a cylinder, use the formula $V = Bh$, where B represents the _____.

Because the base is in the shape of a circle, $B = \pi r^2 =$ _____.

Substitute the values of B and h into the volume formula: $V = Bh = ($_____$)($_____$)$

Simplify the expression to find the volume.

The volume of the cylinder is _____ ft^3.

Lesson Practice

Choose the correct answer.

1. In the popcorn canister below, the diameter is half the height. What is the approximate volume of the canister? Use 3.14 as an approximation for π.

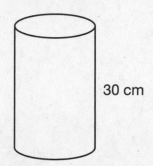

30 cm

   (1) 707 cm³
   (2) 1,413 cm³
   (3) 5,299 cm³
   (4) 21,205 cm³

2. The diameter of a cylinder is doubled, and the height is cut in half. What will happen to the volume of the cylinder?

   (1) The volume will stay the same.
   (2) The volume will be half the original volume.
   (3) The volume will be twice the original volume.
   (4) The volume will be four times the original volume.

3. A tank in the shape of a rectangular prism is emptying water into a cylindrical tank as shown below. Approximately how high will the water rise in the cylindrical tank?

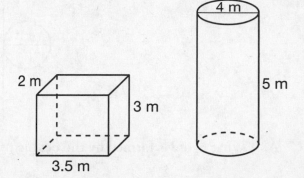

2 m   3 m   4 m 5 m

3.5 m

   (1) 0.4 m
   (2) 1.1 m
   (3) 1.3 m
   (4) 1.7 m

4. The lateral area of a cylinder is in the shape of a

   (1) circle.
   (2) triangle.
   (3) rectangle.
   (4) pentagon.

5. The wheel of a steamroller is in the shape of a cylinder. The length of the wheel is 6 feet, and the diameter of the wheel is 4 feet. To the nearest hundredth, how many square feet will one revolution of the wheel cover?

   (1) 24 ft²
   (2) 48.56 ft²
   (3) 75.36 ft²
   (4) 150.72 ft²

OPEN-ENDED QUESTION

6. Use the net shown below to answer each question.

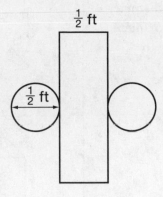

$\frac{1}{2}$ ft

$\frac{1}{2}$ ft

A. What figure is formed by the net pictured above?

_____

B. What is the lateral surface area of the figure rounded to the nearest thousandth?

_____

_____

C. What is the volume of the figure rounded to the nearest thousandth?

_____

_____

_____

Cones

G.G.15

A **cone** is similar to a pyramid, but the base of a cone is a circle. The altitude of a cone is the perpendicular distance between the vertex and the base. The height, h, is the length of the altitude. Also like a pyramid, the volume of a cone can be found by multiplying the product of the area of the base, B, and the altitude by one-third.

> **Volume of a Cone**
> $$V = \frac{1}{3}Bh = \frac{1}{3}\pi r^2 h$$

The diagram below shows that the volume of a cone is one-third the volume of a cylinder having the same base and the same height.

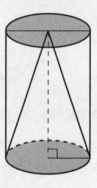

EXAMPLE 1

What is the volume of the cone below?

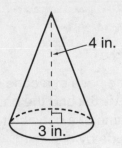

4 in.

3 in.

STRATEGY   **Use the volume formula.**

STEP 1   Find B, the area of the base.

Because the base of a cone is a circle, $B = \pi r^2$.

The diameter of the base is 3 in.; therefore, $r = 1.5$ in.

Let 3.14 be an approximation for π.

$B = \pi r^2$

$B = 3.14(1.5)^2$

$B = 7.065$

STEP 2   Identify the height.

The altitude is the perpendicular height of the cylinder.

height = 4 in.

STEP 3   Substitute the values of B and h into the volume formula and solve.

$V = \frac{1}{3}Bh$

$V = \frac{1}{3}(7.064)(4)$

$V = 9.42$

SOLUTION   **The volume of the cone is 9.42 in.^{3}**

The slant height, *l*, of a cone is the distance from the vertex to a point on the edge of the circular base. To find the lateral area of a cone, multiply the product of the circumference of the base (π*d* or 2π*r*) and the slant height by one-half.

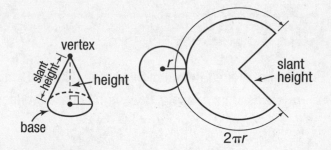

Lateral Area of a Cone

$$LA = \tfrac{1}{2}(2\pi r)l = \pi rl$$

The surface area of a cone is the sum of the base area and the lateral area.

EXAMPLE 2

As part of an art project, Emily is painting the entire surface area of a cone. The cone has a base diameter of 12 inches, a slant height of 10 inches, and an altitude measuring 8 inches. What is the total surface area Emily will be painting?

STRATEGY   **Add the area of the circular base and the lateral surface area.**

STEP 1   Find the area of the base.

Because the base is a circle, $A = \pi r^2$.

The diameter is 12 inches; therefore, the radius is 6 inches.

Let 3.14 be an approximation for π.

$A = 3.14(6^2)$

$A = 113.04$ in.^{2}

STEP 2   Find the lateral surface area.

$LA = \pi rl$

Identify radius *r* and slant height *l*: *r* = 6 and *l* = 10.

$LA = (3.14)(6)(10)$

$LA = 188.4$ in.^{2}

STEP 3   Add the area of the base and the lateral surface area to find the total surface area of the cone.

Total Area $= A + LA$

Total Area $= 113.04 + 188.4 = 301.44$ in.^{2}

SOLUTION   **The surface area to be painted is 301.44 in.^{2}**

COACHED EXAMPLE

A cone-shaped funnel is filled with flour and emptied into a container the shape of a rectangular prism as shown below. How high in the container will the flour reach when it is leveled?

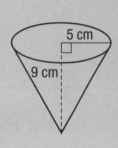
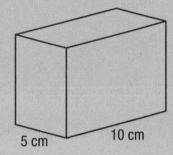

THINKING IT THROUGH

First, find the volume of the flour in the cone: V_{cone} = _____.

The base of a cone is a _____. Then the area of the base B = π(_____).

The radius of the base is _____. Let 3.14 be an approximation for π.

Substitute these values into the formula for the area of the base of the circle and simplify.

The area of the base B = _____.

The length of the _____ of the cone is h. Therefore, h is _____.

Substitute B and h into the volume formula and simplify.

The volume of the cone is _____.

The flour will occupy this same volume in the rectangular prism container.

V_{prism} = _____

Find the area of the base B. Since the base is a rectangle, B = (_____)(_____).

The length of the rectangular base is _____ and the width is _____.
Then B = _____.

You now know the volume of the flour in the prism and the area of the base B. You need to find the height, h, that the flour reaches.

Substitute the known values into the formula for the volume of a prism: _____ = _____h.

Finally, solve for h in this equation.

The height of the flour in the container is _____.

Lesson Practice

Choose the correct answer.

1. A cone has a volume of 600 cm³. If the radius is 10 cm, what is the approximate length of the altitude of the cone?

    (1) 5 cm
    (2) 6 cm
    (3) 50 cm
    (4) 60 cm

2. A metal triangular flag is on a pole as shown in the diagram. On a windy day, the flag spins around the pole so fast that it looks like a three-dimensional shape. Which shape would the spinning flag create?

    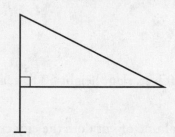

    (1) sphere
    (2) pyramid
    (3) cylinder
    (4) cone

3. The diameter of the base of a cone is quadrupled, and the height remains the same. What will happen to the volume of the cone?

    (1) The volume will stay the same.
    (2) The volume will be twice the original volume.
    (3) The volume will be four times the original volume.
    (4) The volume will be sixteen times the original volume.

4. Find the total surface area of the cone below.

    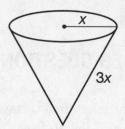

    (1) $4\pi x$
    (2) $4\pi x^2$
    (3) $16\pi x$
    (4) $16\pi x^2$

5. The volumes of cylinder 1 and cylinder 2 are the same. The radius of cylinder 1 is half the radius of cylinder 2. What must be true about the height of the two cylinders?

(1) The height of cylinder 1 is $\frac{1}{2}$ the height of cylinder 2.

(2) The height of cylinder 1 is 2 times the height of cylinder 2.

(3) The height of cylinder 1 is $\frac{1}{4}$ the height of cylinder 2.

(4) The height of cylinder 1 is 4 times the height of cylinder 2.

6. Which figure shown below has the greatest volume?

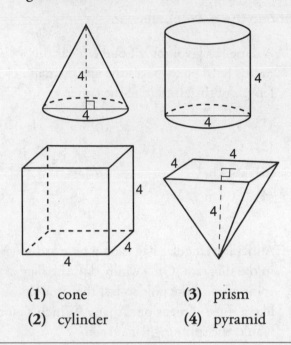

(1) cone   (3) prism

(2) cylinder   (4) pyramid

OPEN-ENDED QUESTION

7. The city stores salt in a cone-shaped tent with base diameter 60 feet, slant height 50 feet, and altitude measuring 40 feet.

A. If half of the volume of the tent is filled with salt, what volume of salt does the city currently have in supply?

_____

_____

B. The exterior of the tent needs to be painted after severe weather has damaged it. What is the surface area of the tent exterior?

_____

_____

C. If one gallon of paint will provide 350 square feet of coverage, how many gallons of paint are needed to paint the entire tent exterior? Give your answer as a whole number.

_____

_____

Spheres

G.G.16

A **sphere** is the set of all points that are a given radius, *r*, from the center, *C*.

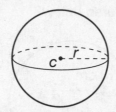

The intersection of a plane and a sphere can be either a single point or a circle. If the intersection contains the center of the sphere, then the intersection creates the *great circle* of the sphere. A **great circle** is the largest circle that can be drawn on a sphere. If two planes intersect a sphere the same distance from its center, then the circles created are congruent.

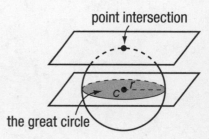

point intersection

the great circle

EXAMPLE 1

Suppose you could travel across the surface of the earth in a great circle. You would begin your trip at the South Pole, travel over the North Pole, and return again to the South Pole. Assume the earth is a perfect sphere with a radius of 3,950 miles. What is the total distance you would travel?

STRATEGY   **Find the circumference of the earth's great circle.**

STEP 1   Identify the radius of the great circle.

Since the great circle shares the center of the sphere, with Earth, they both have the same radius.

The radius, *r*, is 3,950 miles.

STEP 2   Identify the formula and calculate the circumference.

The circumference of the great circle will be the total distance traveled.

The circumference $C = 2\pi r$.

Let 3.14 be the approximation for π.

$C = 2(3.14)(3,950) = 24,806$

SOLUTION   **The total distance traveled is 24,806 miles.**

The surface area of a sphere is four times the product of π and the radius squared.

Surface Area of a Sphere
$SA = 4\pi r^2$

EXAMPLE 2

What is the surface area of the figure pictured below?

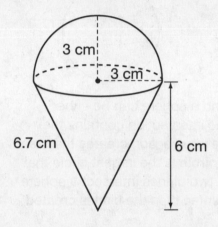

STRATEGY   **Add the lateral surface area of the cone to the surface area of the half-sphere.**

STEP 1   Find half the surface area of the sphere. Let 3.14 be an approximation for π.

Identify the radius: $r = 3$ cm.

$SA = 4\pi r^2$

$SA = 4(3.14)(3^2) = 113.04$

Take half the total surface area to find the surface area of the half-sphere.

$\frac{113.04}{2} = 56.52$ cm^2

STEP 2   Find the lateral surface area of the cone.

Use the formula: $LA = \pi r l$

Identify radius, r, and slant height, l: $r = 3$ and $l = 6.7$.

$LA = (3.14)(3)(6.7)$

$LA = 63.11$ cm^2

STEP 3   Add the two surface areas from steps one and two to find the total surface area of the figure.

Total Area = $56.52 + 63.11 = 119.63$ cm^2

SOLUTION   **The surface area of the figure is 119.63 cm^2.**

The volume of a sphere is four-thirds the product of π and the radius cubed.

> **Volume of a Sphere**
> $$V = \frac{4}{3}\pi r^3$$

EXAMPLE 3

A candle company uses spherical molds to create their best-selling holiday candles. Each finished candle has a diameter of 4 inches. The company will produce 3,000 red candles for their holiday inventory. To the nearest cubic inch, what volume of red wax will the company need to purchase to create the desired amount of stock?

STRATEGY   **Find the volume of wax in one candle and multiply by the number needed.**

STEP 1   Find the volume of wax in one candle.

Identify the radius. Since the diameter is 4 inches, $r = 2$ in.

Let 3.14 be an approximation for π and substitute values into the formula for the volume of a sphere.

$$V = \frac{4}{3}\pi r^3$$

$$V = \frac{4}{3}(3.14)(2^3) = 33.49 \text{ in.}^3$$

STEP 2   Multiply the volume of one candle by the total number of candles needed.

The company wants to produce 3,000 red candles.

$$(33.49)(3,000) = 100,470 \text{ in.}^3$$

SOLUTION   **The company will need 100,470 cubic inches of red wax.**

COACHED EXAMPLE

The surface area of a sphere is 144π m^2. What is the volume of the sphere?

THINKING IT THROUGH

Identify the formula for the surface area of a sphere: $SA =$ _____.

Then, substituting the known surface area into this formula, _____ $=$ _____.

Solve this equation for r.

First, divide both sides by 4.

The simplified equation is _____ $=$ _____.

Now, divide both sides by π.

The simplified equation is _____ $=$ _____.

Finally, to solve for r, take the square root of both sides of the equation.

$r =$ _____.

Identify the formula for the volume of a sphere: $V =$ _____.

Substitute r into this formula and solve for the volume.

_____

_____

The volume of the sphere is _____ **m^3.**

Lesson Practice

Choose the correct answer.

1. Which of the following is not a true statement about a sphere?

   (1) The radius of a great circle is also the radius of the sphere.

   (2) The intersection of a plane and a sphere may be a circle.

   (3) The intersection of a plane and a sphere may be a point.

   (4) Two planes that intersect a sphere equidistant from its center always create parallel circles.

2. To the nearest hundredth, what is the surface area of a sphere with diameter 14 inches?

   (1) 87.92 square inches

   (2) 351.68 square inches

   (3) 615.44 square inches

   (4) 1,436.03 square inches

3. The diameter of a sphere is doubled. What will happen to the volume of the sphere?

   (1) The volume will be twice the original volume.

   (2) The volume will be four times the original volume.

   (3) The volume will be eight times the original volume.

   (4) The volume will be sixteen times the original volume.

4. A spherical ball is placed in a cube-shaped box as pictured below. The sphere touches the edge of the box on all sides. To the nearest cubic centimeter, what is the volume of the empty space in the box?

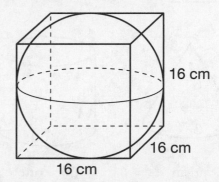

16 cm
16 cm
16 cm

   (1) 856 cubic centimeters

   (2) 1,952 cubic centimeters

   (3) 2,144 cubic centimeters

   (4) 4,096 cubic centimeters

5. The surface area of a sphere is 324π cubic meters. What is the volume of the sphere, in terms of π?

   (1) $9\pi \text{ m}^3$

   (2) $524\pi \text{ m}^3$

   (3) $972\pi \text{ m}^3$

   (4) $1,082\pi \text{ m}^3$

6. Which figure shown below has the greatest volume?

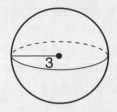

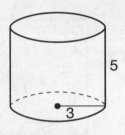

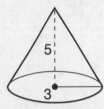

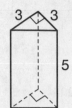

(1) prism **(3)** cone

(2) sphere **(4)** cylinder

7. Which segment is a radius in the sphere below?

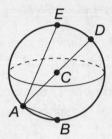

(1) $\overline{AD}$ **(3)** $\overline{AB}$

(2) $\overline{AC}$ **(4)** $\overline{AE}$

OPEN-ENDED QUESTION

8. An ice cream vendor is selling ice cream in the different containers as pictured below.

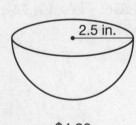

$4.90

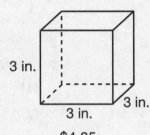

$4.85

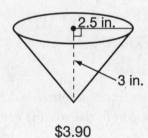

$3.90

A. Which container is the best value? Explain.

_____

_____

B. The half sphere is placed in a cylinder where its top and bottom touch the top and bottom of the cylinder, and its sides touch the sides of the cylinder. What is the volume of the cylinder?

_____

_____

Angles in a Triangle

A **triangle** is a geometric figure with three sides. In a triangle or polygon, the **interior angles** are formed by the sides at each vertex. There are three interior angles in a triangle.

EXAMPLE 1

What is the sum of the measures of the interior angles of a triangle?

STRATEGY   **Draw several types of triangles and measure their angles with a protractor.**

STEP 1   Draw one of each of the following types of triangles: acute, obtuse, right.

STEP 2   Measure and label each interior angle in all three triangles using a protractor.

STEP 3   Find and record the sum of the interior angles of each of the triangles you measured in Step 2. Make a conjecture about the sum of the measures of the interior angles in a triangle.

SOLUTION   **The sum of the interior angle measures of a triangle is equal to 180°.**

EXAMPLE 2

Prove that the sum of the measures of the interior angles of a triangle is equal to 180°.

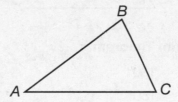

STRATEGY   **Given △ABC, show that m∠1 + m∠2 + m∠3 = 180°.**

STEP 1   Draw line *BD* through the point *B* parallel to $\overline{AC}$. This line is unique according to the Unique Line Postulate. Label the angles numerically.

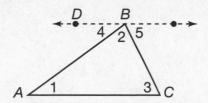

STEP 2   Use the Angle Addition Postulate to write the following equations.

$m\angle DBC + m\angle 5 + 180°$;

$m\angle DBC = m\angle 4 + m\angle 2$

STEP 3   Use the Substitution Property to rewrite the first equation.

$m\angle 4 + m\angle 2 + m\angle 5 = 180°$

STEP 4   Identify pairs of congruent alternate interior angles.

$\angle 4 \cong \angle 1$, or $m\angle 4 = m\angle 1$

$\angle 5 \cong \angle 3$, or $m\angle 5 = m\angle 3$

STEP 5   Use the Substitution Property.

$m\angle 1 + m\angle 2 + m\angle 3 = 180°$

SOLUTION   **The sum of the interior angle measures of a triangle is 180°.**

> **Triangle Sum Theorem:** The sum of the interior angles in a triangle is equal to 180°.

EXAMPLE 3

In the triangle below, what is the $m\angle X$?

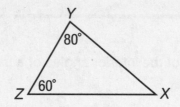

STRATEGY   **Apply the Triangle Sum Theorem.**

$m\angle X + m\angle Y + m\angle Z = 180°$

$m\angle X + 80° + 60° = 180°$

$m\angle X + 140° = 180°$

SOLUTION $m\angle X = 40°$

COACHED EXAMPLE

What is m∠B?

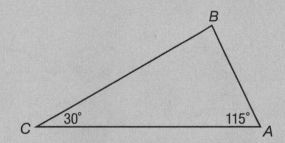

THINKING IT THROUGH

Because of the _____Theorem, m∠A + m∠B + m∠C = _____.

In this equation, replace the angle measures with the values provided in the diagram.

_____ + m∠B + _____ = _____

Add the measures of ∠A and ∠C to get _____.

Subtract the sum from _____ to get m∠ _____.

The measure of angle B is _____.

Lesson Practice

Choose the correct answer.

1. What is the measure of ∠*R*?

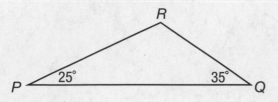

(1) 100°

(2) 110°

(3) 120°

(4) 135°

2. What is the measure of angle *G*?

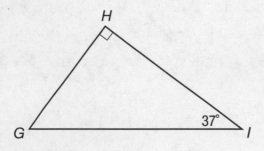

(1) 53°

(2) 63°

(3) 90°

(4) 143°

3. What is the measure of angle *D*?

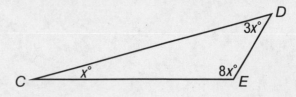

(1) 15°

(2) 45°

(3) 115°

(4) 120°

4. What is the measure of ∠*CAB*?

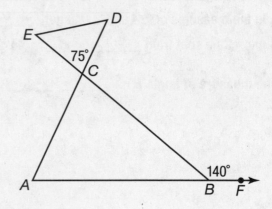

(1) 40°

(2) 65°

(3) 75°

(4) 80°

OPEN-ENDED QUESTION

5. How many angles in a triangle can be greater than 90°? Explain your answer.

_____

_____

_____

_____

Exterior Angle Theorem

An **exterior angle** of a triangle is formed when one side of a triangle is extended. In the diagram, ∠4 is an exterior angle.

The **adjacent interior angle** is the angle of the triangle beside which the side is extended. In the diagram, ∠1 is the adjacent interior angle.

The other two angles of the triangle, 2 and 3, are the **remote interior angles**.

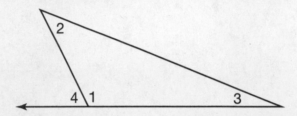

> **Exterior Angle Theorem**: The measure of an exterior angle of a triangle equals the sum of the measures of the two remote interior angles.

The exterior angle and its adjacent interior angle form a linear pair. Because their non common sides are opposite rays, a linear pair forms a line.

> **Linear Pair Postulate:** If two angles form a line, then the sum of these two angles is 180°.

EXAMPLE 1

Prove the Exterior Angle Theorem.

Given: $\triangle ABC$ with exterior angle $\angle BAD$ and remote interior angles, $\angle B$ and $\angle C$.

Prove: $m\angle BAD = m\angle B + m\angle C$

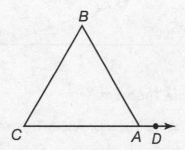

STRATEGY   Write a two-column proof.

Statements	Reasons
1. $\triangle ABC$ with exterior angle $\angle BAD$	1. Given
2. $m\angle BAD + m\angle BAC = 180°$	2. Linear Pair Postulate
3. $m\angle B + m\angle C + m\angle BAC = 180°$	3. Triangle Sum Theorem
4. $m\angle BAD + m\angle BAC = m\angle B + m\angle C + m\angle BAC$	4. Transitive Property
5. $m\angle BAD = m\angle B + m\angle C$	5. Subtraction Property

SOLUTION   The two-column proof is shown above.

EXAMPLE 2

In $\triangle WXY$, $m\angle X = 120°$, and $m\angle WYZ = 4 \cdot m\angle W$. Find $m\angle W$.

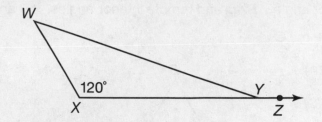

STRATEGY   **Use the Exterior Angle Theorem.**

STEP 1   Write an equation using the Exterior Angle Theorem.

$m\angle WYZ = m\angle X + m\angle W$

STEP 2   Replace $m\angle X$ in the equation with 120°, and replace $m\angle WYZ$ in the equation with $4 \cdot m\angle W$.

$4 \cdot m\angle W = 120° + m\angle W$

STEP 3   Solve for $m\angle W$.

$4 \cdot m\angle W = 120° + m\angle W$

$3 \cdot m\angle W = 120°$ Subtract $m\angle W$ from both sides.

$m\angle W = 40°$ Divide both sides by 3.

SOLUTION   **The measure of $\angle W = 40°$.**

Because an exterior angle is equal to the sum of the remote interior angles, it must be greater than either one of them. This is the Exterior Angle Inequality Theorem.

> **Exterior Angle Inequality Theorem:** The measure of an exterior angle of a triangle is greater than the measure of either remote interior angle.

EXAMPLE 3

Prove the Exterior Angle Inequality Theorem.

Given: ∠1 is an exterior angle of △*DEF*.

Prove: m∠1 > m∠*D*; m∠1 > m∠*E*

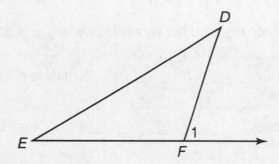

STRATEGY   **Write a paragraph proof.**

It is given that ∠1 is an exterior angle, so, by the Exterior Angle Theorem, m∠1 = m∠*D* + m∠*E*. Then, using the Partition Postulate (Lesson 7), because the whole is equal to the sum of its parts, the whole must be larger than each individual part. Therefore, m∠1 > m∠*D*; m∠1 > m∠*E*.

SOLUTION   **The paragraph proof is shown above.**

COACHED EXAMPLE

Find the value of *x*.

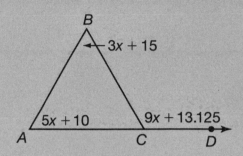

THINKING IT THROUGH

Because you are given an exterior angle and two remote interior angles, a good theorem to use here is the _____.

Set the sum of the measures of ∠ _____ and ∠*B* equal to m∠*BCD*.

Write the equation. _____

Combine like terms, then subtract _____ from both sides of the equation.

Solve for *x*. _____

x = _____.

Lesson Practice

Choose the correct answer.

In Exercises 1–3, use the given information to find the measure of the specified angle. The diagram is not drawn to scale.

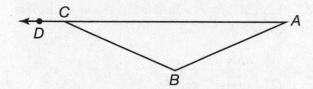

1. Find m∠*BCD*, given m∠*A* =37° and m∠*B* = 45°.

   (1) 37°

   (2) 45°

   (3) 82°

   (4) 98°

2. Find m∠*A*, if m∠*B* = 110° and m∠*BCD* = 135°.

   (1) 15°

   (2) 25°

   (3) 100°

   (4) 110°

3. Find m∠*B*, if m∠*A* = 3*x* and m∠*BCD* = 9*x* + 7.

   (1) 12*x*

   (2) 12*x* + 7

   (3) 6*x*

   (4) 6*x* + 7

Use this diagram for Exercises 4–6.

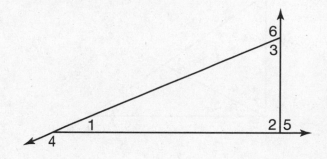

4. If m∠1 = *x*, m∠3 = 2*x* + 11, and m∠5 = 4*x* − 14, find the value of *x*.

   (1) 25°

   (2) 30°

   (3) 61°

   (4) 86°

5. Find m∠6 if m∠1 = 30° and m∠2 = 85°.

   (1) 150°

   (2) 115°

   (3) 95°

   (4) 55°

6. Find m∠4 + m∠5 + m∠6.

   (1) 180°

   (2) 360°

   (3) 540°

   (4) 720°

Use the diagram for Exercises 7–8.

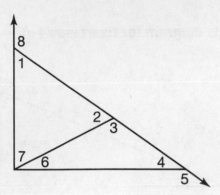

7. Find all angles whose measures must be greater than ∠1.

   (1) only ∠2

   (2) only ∠3

   (3) angles 8 and 5

   (4) angles 3, 4, and 5

8. Find all angles whose measures must be less than ∠8.

   (1) only ∠2

   (2) angles 2 and 7

   (3) angles 2, 4, and 7

   (4) angles 2, 4, 6, and 7

OPEN-ENDED QUESTION

9. Find the measure of the base angles of an isosceles △*ABC*, given the exterior angle of its vertex angle has a measure of 140°. Explain the strategy you used to solve this problem.

_____

_____

_____

18 Triangle Inequality Theorem

Although there are an infinite number of different triangles, not every set of three lengths will form a triangle. There is a relationship between the sum of the lengths of two sides of a triangle and the length of the third side.

EXAMPLE 1

Make a conjecture about the sum of the lengths of any two sides of a triangle.

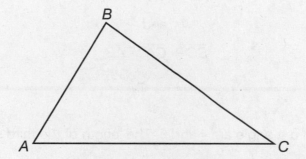

STRATEGY   **Examine the relationship between the lengths of sides.**

     STEP 1   Draw several triangles.

     STEP 2   Measure and record the lengths of the sides.

     STEP 3   Add the lengths of two sides of a triangle.

                       Compare the sum of the two sides to the length of the third side. Repeat for the sum of the lengths of each pair of sides.

     STEP 4   Make a conjecture about the relationship between the sum and the third side.

SOLUTION   **Conjecture: The sum of the lengths of any two sides of a triangle is greater than the length of the third side.**

> **Triangle Inequality Theorem**: The sum of the lengths of any two sides of a triangle is greater than the length of the third side.

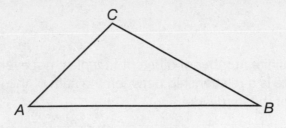

In $\triangle ABC$, all of the following are true:

$$AB + BC > CA,$$

$$AB + CA > BC,$$

and

$$\overline{BC} + \overline{CA} > \overline{AB}.$$

EXAMPLE 2

The lengths of two sides of a triangle are 4 and 6. The length of the third side must be greater than _____, but less than _____.

STRATEGY   Identify the range of possible values.

STEP 1   Let x be the length of the third side.

STEP 2   Write the inequalities. By the Triangle Inequality Theorem:

$x + 4 > 6$ $4 + 6 > x$ $x + 6 > 4$

STEP 3   Solve each inequality.

$x > 2$ $10 > x$ $x > -2$

SOLUTION   **The length of the third side must be greater than 2, but less than 10.**

Note: The inequality $x + 6 > 4$ did not provide a solution. In the inequality $x + 6 > 4$, the sum of any positive number and 6 would be greater than 4. Therefore to solve the inequality the x value would need to be a negative number. Because x is supposed to represent a length, it has to be a positive number.

EXAMPLE 3

Is it possible for a triangle to have sides with the lengths 18, 12, and 6?

STRATEGY   **Apply the Triangle Inequality Theorem.**

STEP 1   Write inequalities for each pair of sides.

$18 + 12 \overset{?}{>} 6$ $18 + 6 \overset{?}{>} 12$ $12 + 6 \overset{?}{>} 18$

STEP 2   Determine if any inequalities fail to hold true.

$30 > 6$ $24 > 12$ $18 \not> 18$

SOLUTION   **It is not possible for a triangle to have the given lengths because they do not satisfy the Triangle Inequality Theorem.**

COACHED EXAMPLE

Johannes has three pieces of fencing that have lengths of 9 ft, 6.5 ft, and 7.25 ft. He wants to use the three pieces of wood fencing to make a triangular pen for his rabbits. Is it possible to use the pieces of fencing to create a triangular pen without having to cut any of them shorter? Explain.

THINKING IT THROUGH

The Triangle _____ Theorem states the sum of the lengths of any two sides of a triangle is _____ than the length of the other side.

Check the following:

$9 + 6.5 \overset{?}{>}$ _____   yes

$9 +$ _____ $\overset{?}{>} 6.5$ yes

$6.5 + 7.25 \overset{?}{>} 9$ _____

Because each of the inequalities above _____ the requirements of the Triangle Inequality Theorem, Johannes _____

_____ .

Lesson Practice

Choose the correct answer.

In Exercises 1–3, determine whether it is possible for a triangle to have sides with the given lengths.

1. 17, 12, and 5

   (1) yes
   (2) no; $17 + 12 \not> 5$
   (3) no; $12 + 5 \not> 17$
   (4) no; $17 + 5 > 12$

2. 10, 6, and 5

   (1) yes
   (2) no; $6 + 5 > 11$
   (3) no; $10 + 5 \not> 6$
   (4) no; $10 + 6 \not> 5$

3. 4.8, 2.7, and 2.2

   (1) yes
   (2) no; $4.8 + 2.7 \not> 2.2$
   (3) no; $4.8 + 2.2 > 2.7$
   (4) no; $2.7 + 2.2 \not> 4.8$

In Exercises 4–6, given two sides of a triangle, find a range of possible values for the length of the third side.

4. $AB = 6$, $BC = 4$

   (1) $-2 < CA < 10$
   (2) $2 < CA < 10$
   (3) $2 \leq CA \leq 10$
   (4) $-2 \leq CA \leq 10$

5. $XY = 6$, $YZ = 9$

   (1) ZX is between -3 and 3.
   (2) ZX is between -3 and 15.
   (3) ZX is between 3 and 15.
   (4) ZX is more than 15.

6. $PQ = 3.6$, $RQ = 7.1$

   (1) RP is between 3.5 and 10.7.
   (2) RP is between 3.4 and 10.8.
   (3) RP is between -3.5 and 10.7.
   (4) RP is less than 3.5 and more than 10.7.

7. The sides of a parallelogram have lengths 9 and 12.

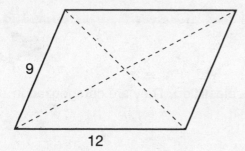

(1) The diagonals are between 15 and 20.

(2) The diagonals are 20.

(3) The diagonals are greater than 3.

(4) The diagonals are between 3 and 21.

8. If $\triangle LMN$ were drawn to scale, which side would be the longest?

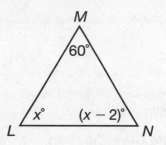

(1) $\overline{LM}$

(2) $\overline{MN}$

(3) $\overline{NL}$

(4) All sides of the triangle have the same length.

OPEN-ENDED QUESTION

9. The distance between Niagara Falls and Syracuse is 165 miles. The distance between Syracuse and New York City is 193 miles. Find a range of values for the distance between Niagara Falls and New York City. Show your work.

_____

_____

_____

_____

Relative Size in Triangles

G.G.34

There are two different ways in which triangles can be classified. They are categorized in the outline below:

Triangle Classification

I. By number of congruent sides

    A. **Scalene** – no congruent sides

    B. **Isosceles** – two congruent sides

    C. **Equilateral** – all sides congruent

II. By relative size of angles

    A. **Acute** – three acute angles

    B. **Obtuse** – one obtuse angle

    C. **Right** – one right angle

    D. **Equiangular** – all angles congruent

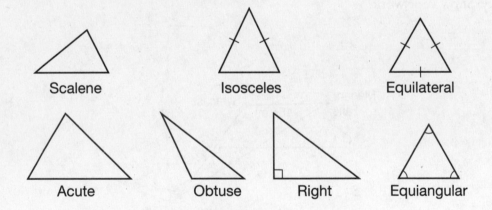

EXAMPLE 1

How do the sides and angles of a triangle relate to one another?

STRATEGY   **Draw several different scalene triangles (they can be acute, obtuse, or right). Measure the sides with a ruler, and measure the angles with a protractor.**

STEP 1   Draw three scalene triangles; they can be acute, obtuse, or right.

STEP 2   Measure and record each of the three sides in each of the triangles you drew in Step 1.

STEP 3   Measure and label each of the three angles in each of the triangles you drew in Step 1.

STEP 4   Compare the angles opposite the longest sides of your triangles and the angles opposite the shortest sides of your triangles.

STEP 5   Compare the sides opposite the largest angles of your triangles and the sides opposite the smallest angles of your triangles.

SOLUTION   **Make a conjecture that the angles opposite the longer sides are greater in measure, and the sides opposite the larger angles are greater in length.**

These angle-side relationships are further explained in the following two theorems.

Unequal Sides Theorem: If two sides of a triangle are unequal in length, then the measure of the angle opposite the longer side is greater than the measure of the angle opposite the shorter side.

Unequal Angles Theorem: If angles of a triangle are unequal in measure, then the side opposite the larger angle is longer than the side opposite the smaller angle.

EXAMPLE 2

Identify the largest and smallest angles in the given triangle.

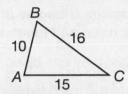

STRATEGY   Given △*ABC*, identify its largest and smallest angles.

   STEP 1   Because of the Unequal Sides Theorem, since $\overline{BC}$ is the longest side, ∠*A*, the angle opposite it, is the largest angle.

   STEP 2   Also by the Unequal Sides Theorem, since $\overline{AB}$ is the shortest side, ∠*C*, the angle opposite it, is the smallest angle.

SOLUTION   In △*ABC*, the largest angle is ∠*A*, and the smallest angle is ∠*C*.

COACHED EXAMPLE

Name the longest and shortest sides in △*RST*.

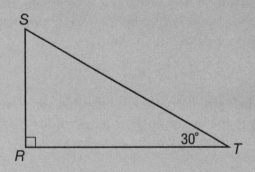

THINKING IT THROUGH

Because of the _____ Theorem, we know the measure of

∠*S* is _____ than ∠*T*, since ∠*S* + ∠*T* = _____°.

∠*T* is the _____ angle of △*RST*.

By the _____ Angles Theorem, because ∠*R* is the largest angle, $\overline{ST}$,
the _____ side, is the longest side.

By the Unequal _____ Theorem, because _____ is the
smallest angle, $\overline{RS}$, the opposite side, is the _____ side.

Lesson Practice

Choose the correct answer.

In Exercises 1–4, given the sides, identify the largest and smallest angle (in that order) in each triangle.

1. △DEF; DE = 3, EF = 9, FD = 7

    (1) ∠E, ∠F
    (2) ∠D, ∠F
    (3) ∠D, ∠E
    (4) ∠E, ∠D

2. △PQR; PQ = 12, QR = 9, RP = 11.5

    (1) ∠R, ∠P
    (2) ∠P, ∠R
    (3) ∠R, ∠Q
    (4) ∠P, ∠Q

3. △ABC; AB = 7.8, BC = 8.7, CA = 9.3

    (1) ∠B, ∠A
    (2) ∠A, ∠B
    (3) ∠A, ∠C
    (4) ∠B, ∠C

4. △XYZ; XY = 0.9, YZ = 0.7, ZX = 0.75

    (1) ∠Z, ∠Y
    (2) ∠Z, ∠X
    (3) ∠Y, ∠X
    (4) ∠X, ∠Z

In Exercises 5–8, given two angles, identify the longest and shortest side (in that order) in each triangle.

5. △JKL; m∠J = 40°, m∠K = 55°

    (1) $\overline{LJ}$, $\overline{KL}$
    (2) $\overline{JK}$, $\overline{LJ}$
    (3) $\overline{JK}$, $\overline{KL}$
    (4) $\overline{LJ}$, $\overline{JK}$

6. △STU; m∠T = 28°, m∠U = 15°

    (1) $\overline{TU}$, $\overline{ST}$
    (2) $\overline{ST}$, $\overline{TU}$
    (3) $\overline{TU}$, $\overline{US}$
    (4) $\overline{US}$, $\overline{ST}$

7. △MAT; m∠A = 65°, m∠T = 64°

    (1) $\overline{MA}$, $\overline{AT}$
    (2) $\overline{MT}$, $\overline{MA}$
    (3) $\overline{MT}$, $\overline{AT}$
    (4) $\overline{AT}$, $\overline{MT}$

8. △CVW; m∠W = 90°, m∠C = 46°

    (1) $\overline{CV}$, $\overline{VW}$
    (2) $\overline{VW}$, $\overline{CV}$
    (3) $\overline{WC}$, $\overline{CV}$
    (4) $\overline{CV}$, $\overline{WC}$

OPEN-ENDED QUESTION

9. In the preceding exercises, were any of the triangles isosceles or equilateral? Explain your answer.

_____

_____

_____

_____

_____

_____

_____

Congruence of Two Triangles

 G.G.28, G.G.29

Two figures with the same size and shape are **congruent,** ≅. In the diagram below, triangles *ABC* and *DEF* are congruent (△*ABC* ≅ △*DEF*). Two triangles are congruent if and only if the **corresponding parts** (angles and sides) of the triangles are congruent. Triangles do not need to have the same orientation to be congruent. The diagram shows the method of labeling the corresponding congruent sides and angles of congruent triangles.

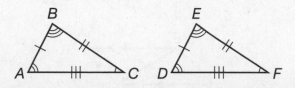

Corresponding and Congruent Angles	Corresponding and Congruent Sides
∠*A* ≅ ∠*D*	$\overline{AB} \cong \overline{DE}$
∠*B* ≅ ∠*E*	$\overline{BC} \cong \overline{EF}$
∠*C* ≅ ∠*F*	$\overline{AC} \cong \overline{DF}$

Although the definition of congruent triangles requires that all six pairs of corresponding parts be congruent, it is not necessary to show all pairs in order to prove that triangles are congruent. The following theorems allow triangles to be proved congruent with only two or three congruent pairs of corresponding parts.

Triangle Congruence

Side-Side-Side (SSS) Congruence Postulate: If three sides of one triangle are congruent to three sides of another triangle, then the triangles are congruent.

Side-Angle-Side (SAS) Congruence Postulate: If two sides and the included angle of one triangle are congruent to two sides and the included angle of another triangle, then the triangles are congruent.

Angle-Side-Angle (ASA) Congruence Postulate: If two angles and the included side of one triangle are congruent to two angles and the included side of another triangle, then the triangles are congruent.

Angle-Angle-Side (AAS) Congruence Theorem: If two angles and a non-included side of one triangle are congruent to two angles and the corresponding non-included side of another triangle, then the triangles are congruent.

Hypotenuse-Leg (HL) Congruence Theorem: If the hypotenuse and one leg of a right triangle are congruent to the hypotenuse and one leg of another right triangle, then the triangles are congruent.

CPCTC: Corresponding parts of congruent triangles are congruent.

EXAMPLE 1

Given: $\overline{EX} \parallel \overline{DF}$

$\overline{DE} \parallel \overline{FX}$

Prove: $\triangle DEF \cong \triangle XFE$

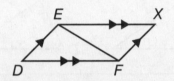

STRATEGY   Write a two-column proof.

Statements	Reasons
1. $\overline{EX} \parallel \overline{DF}$ $\overline{DE} \parallel \overline{FX}$	1. Given
2. $\angle XEF \cong \angle DFE$ $\angle XFE \cong \angle DEF$	2. Alternate Interior Angles Theorem
3. $\overline{EF} \cong \overline{FE}$	3. Reflexive Property
4. $\triangle DEF \cong \triangle XFE$	4. ASA Congruence Postulate

SOLUTION   The two-column proof is shown above.

EXAMPLE 2

Given: $\overline{KA} \cong \overline{DA}$ and $\overline{AT} \cong \overline{AQ}$

Prove: $\triangle KAT \cong \triangle DAQ$

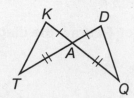

STRATEGY   Write a two-column proof.

Statements	Reasons
1. $\overline{KA} \cong \overline{DA}$ and $\overline{AT} \cong \overline{AQ}$	1. Given
2. $\angle KAT \cong \angle DAQ$	2. Vertical angles
3. $\triangle KAT \cong \triangle DAQ$	3. SAS Congruence Postulate

SOLUTION   The two-column proof is shown above.

EXAMPLE 3

Given: $\overline{AB} \parallel \overline{DC}$, $\overline{AD} \parallel \overline{BC}$

Prove: $\triangle ABD \cong \triangle CDB$.

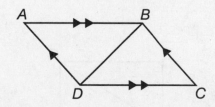

STRATEGY   Write a two-column proof.

Statements	Reasons
1. $\overline{AB} \parallel \overline{DC}$ and $\overline{AD} \parallel \overline{BC}$	1. Given
2. $\angle ABD \cong \angle CDB$ and $\angle ADB \cong \angle CBD$	2. Alternate Interior Angles Theorem
3. $\overline{BD} \cong \overline{DB}$	3. Reflexive Property
4. $\triangle ABD \cong \triangle CDB$	4. ASA Congruence Postulate

SOLUTION   The two-column proof is shown above.

EXAMPLE 4

Given: $\overline{AC} \cong \overline{AE}$; $\angle B \cong \angle D$

Prove: $\triangle ABC \cong \triangle ADE$

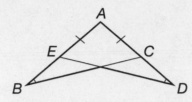

STRATEGY   Write a two-column proof. Redraw the diagram as two separate triangles, as shown.

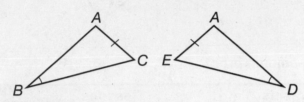

Statements	Reasons
1. $\overline{AC} \cong \overline{AE}$ and $\angle B \cong \angle D$	1. Given
2. $\angle A \cong \angle A$	2. Reflexive Property
3. $\triangle ABC \cong \triangle ADE$	3. AAS Congruence Theorem

SOLUTION   The two-column proof is shown above.

EXAMPLE 5

Given: $\overline{AB} \cong \overline{DC}$; $\angle B \cong \angle D$

Prove: $\triangle ABC \cong \triangle CDA$

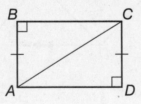

STRATEGY   Write a paragraph proof.

Given $\overline{AB} \cong \overline{CD}$, then show that $\triangle ABC \cong \triangle CDA$. Due to the definition of a right triangle, both triangles ABC and CDA are right triangles. Due to the definition of hypotenuse, $\overline{AC}$ is the hypotenuse of $\triangle ABC$, and $\overline{CA}$ is the hypotenuse of $\triangle CDA$. Due to the Reflexive Property, we can state $\overline{AC} \cong \overline{CA}$. Due to the Hypotenuse-Leg Theorem, $\triangle ABC \cong \triangle CDA$.

SOLUTION   The statement $\triangle ABC \cong \triangle CDA$ is true.

COACHED EXAMPLE

Given: $\angle WYX \cong \angle WYZ$ and $\overline{XY} \cong \overline{ZY}$

Prove: $\triangle WXY \cong \triangle WZY$

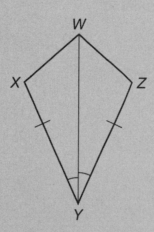

THINKING IT THROUGH

It is given that these two triangles have one set of _____ congruent, and one set of _____ congruent.

The _____ Theorem cannot be used, because these two triangles are not right triangles.

At minimum, show at least one other side or _____ congruent, in order to apply one of the triangle congruence postulates or theorems.

The two triangles have a side that they share in common, which is segment _____.

This segment or side is congruent to itself because of the _____ Property.

There are two _____ congruent and one included _____ congruent, which means we can show triangles *WXY* and *WZY* are congruent by the _____ _____ Postulate.

Lesson Practice

Choose the correct answer.

In Exercises 1–3, state which method(s) can be used to prove the triangles congruent. If no method applies, choose none.

1.

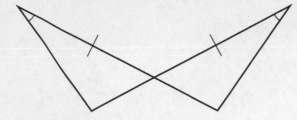

(1) AAS

(2) ASA

(3) ASA and AAS

(4) none

2.

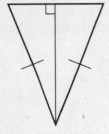

(1) HL

(2) AAS

(3) HL and AAS

(4) none

3.

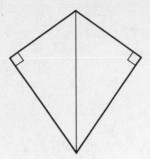

(1) HL

(2) ASA

(3) HL and ASA

(4) none

4. $\triangle ABC \cong \triangle DEF$. What angle corresponds to $\angle CAB$?

(1) $\angle ABC$

(2) $\angle FDE$

(3) $\angle DEF$

(4) $\angle FED$

In Exercises 5–7, choose the answer that completes the proof.

Given: $\overline{AB} \cong \overline{BC}$; $\overline{AT} \cong \overline{BX}$; $\overline{AT} \parallel \overline{BX}$

Prove: $\triangle ABT \cong \triangle BCX$

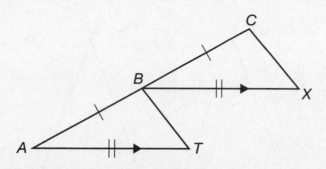

Statements	Reasons
1. $\overline{AB} \cong \overline{BC}$; $\overline{AT} \cong \overline{BX}$; $\overline{AT} \parallel \overline{BX}$	1. Given
2. $\angle CBX \cong \angle$_____	2. _____
3. $\triangle ABT \cong \triangle BCX$	3. _____

5. The missing part of statement 2 is

   (1) *CXB.*

   (2) *BTA.*

   (3) *BAT.*

   (4) *ABT.*

6. The missing reason of step 2 is

   (1) Alternate Interior Angles Theorem.

   (2) Corresponding Angles Theorem.

   (3) Alternate Exterior Angles Theorem.

   (4) the Reflexive Property.

7. The missing reason of step 3 is

   (1) SSS Congruency Postulate.

   (2) AAS Congruency Postulate.

   (3) ASA Congruency Postulate.

   (4) SAS Congruency Postulate.

OPEN-ENDED QUESTION

8. Can these two triangles be proven congruent? Explain your answer.

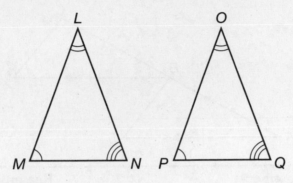

_____

_____

_____

_____

_____

21 Isosceles Triangle Theorem

G.G.31

An **isosceles triangle** is a triangle with at least two congruent sides called **legs**. In an isosceles triangle, the angle formed by the legs is the **vertex angle**. The side opposite this angle is the **base**. The **base angles** of an isosceles triangle are the angles formed by each leg and the base of the triangle.

> **Isosceles Triangle Theorem:** If a triangle is isosceles, then the base angles are congruent.

EXAMPLE 1

Prove the Isosceles Triangle Theorem.

Given: $\overline{AB} \cong \overline{AC}$

Prove: $\angle B \cong \angle C$

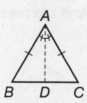

STRATEGY   Write a two-column proof.

Statements	Reasons
1. $\overline{AB} \cong \overline{AC}$	1. Given
2. Draw a bisector of $\angle A$ to the base.	2. Construction of angle bisector
3. $\angle BAD \cong \angle CAD$	3. Definition of angle bisector
4. $\overline{AD} \cong \overline{AD}$	4. Reflexive Property
5. $\triangle ABD \cong \triangle ACD$	5. SAS Congruence Theorem
6. $\angle B \cong \angle C$	6. CPCTC

SOLUTION   The two-column proof is shown above.

EXAMPLE 2

$\triangle XYZ$ is isosceles with base $\overline{XZ}$. Given the measure of $\angle X = 50°$, find the measures of the other two angles.

STRATEGY   **Draw a picture and use the Isosceles Triangle Theorem.**

STEP 1   Draw isosceles $\triangle XYZ$, having $\overline{XZ}$ as its base and $\overline{XY}$ and $\overline{ZY}$ as its legs.

STEP 2   Label $\angle X$ as 50°.

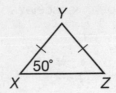

STEP 3   Find $m\angle Z$.

The Isosceles Triangle Theorem confirms the measure of $\angle Z = 50°$.

STEP 4   Use the Triangle Sum Theorem to find the measure of $\angle Y$.

$$m\angle Y = 180° - (2 \cdot 50°)$$
$$= 180° - 100°$$
$$= 80°.$$

SOLUTION   The measure of $\angle Z = 50°$, and the measure of $\angle Y = 80°$.

EXAMPLE 3

Prove the converse of the Isosceles Triangle Theorem: If two angles of a triangle are congruent, then the opposite sides are congruent.

Given: $\angle B \cong \angle C$

Prove: $\overline{AB} \cong \overline{AC}$

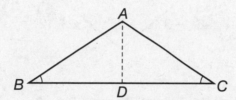

STRATEGY   Write a two-column proof.

Statements	Reasons
1. $\angle B \cong \angle C$	1. Given
2. Draw a bisector of $\angle A$ and extend it to the base.	2. Construction of angle bisector
3. $\angle BAD \cong \angle CAD$	3. Definition of angle bisector
4. $\overline{AD} \cong \overline{AD}$	4. Reflexive Property
5. $\triangle ABD \cong \triangle ACD$	5. AAS Congruence Theorem
6. $\overline{AB} \cong \overline{AC}$	6. CPCTC

SOLUTION   The two-column proof is shown above.

COACHED EXAMPLE

Find the value of *x*.

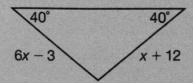

THINKING IT THROUGH

Because of the converse of the Isosceles Triangle Theorem, $6x - 3 =$ _____.

After writing the equation, solve for _____ by subtracting it from both sides.

Then isolate the term 5*x* by _____ 3 to both sides.

Then divide both sides of the equation by the coefficient of *x*, which is _____.

$x =$ _____

Lesson Practice

Choose the correct answer.

In Exercises 1–3, find the measure of each base angle of an isosceles triangle that has a vertex angle with the given measure.

1. 36°

   (1) 64°

   (2) 72°

   (3) 100°

   (4) 144°

2. 50°

   (1) 30°

   (2) 45°

   (3) 65°

   (4) 90°

3. $x°$

   (1) $180 - x$

   (2) $180 - 2x$

   (3) $(180 - x) \div 2$

   (4) $90 - x$

4. In $\triangle STU$, $\angle S \cong \angle U$, $ST = 4x$, $TU = 9x - 10$, and $SU = 7x$. Find ST, TU, and SU.

   (1) 2, 2, 14

   (2) 2, 8, 14

   (3) 8, 8, 14

   (4) 8, 8, 7

In Exercises 5–6, find the measure of the vertex angle of an isosceles triangle that has a base angle with the given measure.

5. 74°

   (1) 32°

   (2) 53°

   (3) 74°

   (4) 106°

6. 35°

   (1) 35°

   (2) 70°

   (3) 110°

   (4) 120°

7. Using the diagram provided, find the value of *x*.

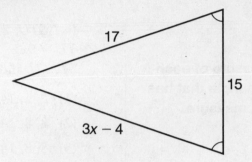

(1) 5

(2) 7

(3) 15

(4) 17

OPEN-ENDED QUESTION

8. Prove that the perpendicular bisector of the base of an isosceles triangle bisects the vertex angle.

22 Concurrency in a Triangle

G.G 21

There are four special segments in a triangle.

The **altitude** of a triangle is the perpendicular segment from a vertex to the opposite side.

A **median** connects a vertex to the midpoint of the opposite side.

Angle bisectors cut each angle into two congruent angles.

A **perpendicular bisector** is the segment that is perpendicular to a side at its midpoint.

Altitude

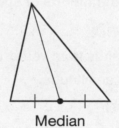

Median

Angle Bisector

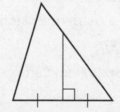

Perpendicular Bisector

EXAMPLE 1

Name each of the four special segments in the diagram.

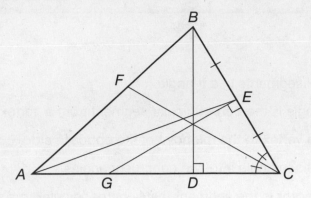

STRATEGY Use the definitions of special segments.

STEP 1 Examine $\overline{BD}$.

It extends from vertex B and is perpendicular to $\overline{AC}$.

$\overline{BD}$ is an altitude.

STEP 2 Examine $\overline{GE}$.

It is perpendicular to $\overline{BC}$ and passes through the midpoint of the side.

$\overline{GE}$ is a perpendicular bisector.

STEP 3 Examine $\overline{AE}$.

It extends from vertex A to the midpoint of $\overline{BC}$.

$\overline{AE}$ is a median.

STEP 4 Examine $\overline{CF}$.

It cuts vertex C into two congruent angles.

$\overline{CF}$ is an angle bisector.

SOLUTION The segments are identified above.

A point is **equidistant** from two points if it is the same distance from each point. Perpendicular bisectors have the characteristic that every point on them is equidistant from the line segment they bisect. Angle bisectors have the characteristic that every point on them is equidistant from the sides of the angle they bisect.

> **Perpendicular Bisector Theorem:** A point is on the perpendicular bisector of a segment if and only if it is equidistant from the endpoints of the segment.
>
> **Converse of the Perpendicular Bisector Theorem:** A point is equidistant from the endpoints of a segment if and only if it is on the perpendicular bisector.
>
> **Angle Bisector Theorem:** A point is on the bisector of an angle if and only if it is equidistant from the sides of the angle.
>
> **Converse of Angle Bisector Theorem:** A point is in the interior of an angle and equidistant from the sides of the angle if and only if it lies on the bisector of the angle.

EXAMPLE 2

Given: $\overline{XZ}$ is the perpendicular bisector of $\overline{WY}$.

Prove: $\triangle WXZ \cong \triangle YXZ$

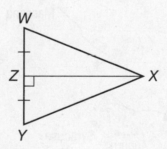

STRATEGY Write a two-column proof.

Statements	Reasons
1. $\overline{XZ}$ is the perpendicular bisector of $\overline{WY}$.	1. Given
2. $\overline{WX} \cong \overline{YX}$	2. Perpendicular Bisector Theorem
3. $\overline{WZ} \cong \overline{YZ}$	3. Definition of perpendicular bisector
4. $\triangle WXZ \cong \triangle YXZ$	4. HL

SOLUTION The two-column proof is shown above.

Concurrent lines are three or more lines that intersect at the same point. Each of the special segments of a triangle are concurrent at a point with specific characteristics. The point of concurrency of the perpendicular bisectors of a triangle is the **circumcenter** of the triangle. The circumcenter is the center of a circle **circumscribed** (drawn around) the triangle. The point is equidistant from the vertices of the triangle, so lines drawn from P to the vertices are radiuses of the circle.

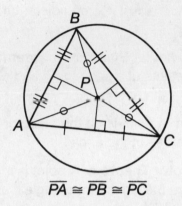

$$\overline{PA} \cong \overline{PB} \cong \overline{PC}$$

EXAMPLE 3

State planners in New York State want to build a tourism bureau in the center of the state. They choose the cities of Massena, Jamestown and Newburgh as outposts of the state. In which city should they build the tourism bureau?

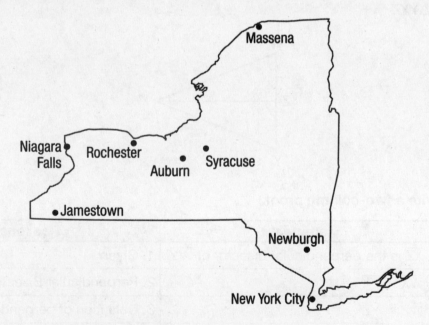

STRATEGY Use the perpendicular bisectors.

STEP 1 Draw segments to connect the three cities and form a triangle.

STEP 2 Measure the length of each segment and mark the midpoints.

STEP 3   Draw the perpendicular bisectors of each side.

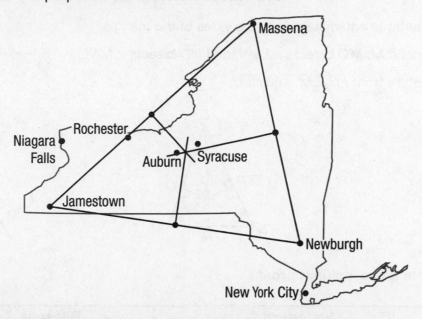

STEP 4   Identify the city closest to the circumcenter of the triangle.

SOLUTION   **The planners should build the tourism center in Auburn.**

The point of concurrency of the angle bisectors of a triangle is the **incenter**. The incenter is the center of an inscribed circle. It is equidistant from the sides of the triangle.

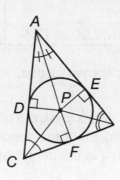

$$DP \cong EP \cong FP$$

EXAMPLE 4

Prove that the incenter is equidistant from the sides of the triangle.

Given: $\overrightarrow{LO}$ bisects $\angle NLM$; $\overrightarrow{MO}$ bisects $\angle LMN$; and $\overrightarrow{NO}$ bisects $\angle MNL$.

Prove: O is equidistant from $\overline{LN}$, $\overline{LM}$, and $\overline{NM}$.

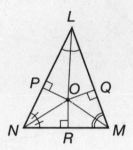

STRATEGY   Write a two-column proof.

Statements	Reasons
1. $\overrightarrow{LO}$ bisects $\angle NLM$; $\overrightarrow{MO}$ bisects $\angle LMN$; and $\overrightarrow{NO}$ bisects $\angle MNL$.	1. Given
2. $\overline{PO} \cong \overline{QO}$	2. Angle Bisector Theorem
3. $\overline{QO} \cong \overline{RO}$	3. Angle Bisector Theorem
4. $\overline{RO} \cong \overline{PO}$	4. Angle Bisector Theorem
5. $\overline{PO} \cong \overline{QO} \cong \overline{RO}$	5. Transitive Property
6. O is equidistant from $\overline{LN}$, $\overline{LM}$, and $\overline{NM}$.	6. Definition of equidistant

SOLUTION   **The two-column proof is shown above.**

The medians of a triangle are concurrent at the **centroid** of the triangle. The centroid is two-thirds of the distance from the vertex to the midpoint of the opposite side. It divides the median into segments whose lengths have a ratio 2:1. It is the center of gravity of the triangle.

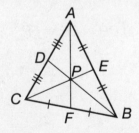

$$AP = \tfrac{2}{3} AF \qquad AP:PF = 2:1$$

$$CP = \tfrac{2}{3} CE \qquad CP:PE = 2:1$$

$$BP = \tfrac{2}{3} BD \qquad BP:PD = 2:1$$

EXAMPLE 5

P is the centroid of △*XYZ*. *PZ* = 15. Find *ZU*. What is the ratio of *PZ* to *PU*?

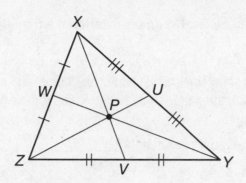

STRATEGY   Use the property of the centroid.

STEP 1   Write an equation using the property of the centroid.

$$PZ = \frac{2}{3} UZ$$

STEP 2   Substitute the known value and solve for *UZ*.

$$15 = \frac{2}{3} UZ$$

$$\frac{3}{2} \cdot 15 = \frac{3}{2}\left(\frac{2}{3}\right) UZ \quad \text{Multiply by the reciprocal of } \frac{2}{3}.$$

$$22.5 = UZ$$

STEP 3   Find the ratio of *PZ* to *PU*.

$$PU = 22.5 - 15 = 7.5$$

$$PZ : PU = 15 : 7.5 = 2 : 1$$

SOLUTION   **The length of *UZ* is 22.5 units. The ratio of the lengths of *PZ* to *PU* is 2:1.**

The altitudes of a triangle are concurrent at a point called the **orthocenter**.

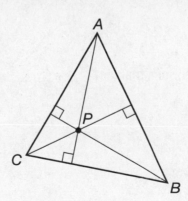

COACHED EXAMPLE

Which points of concurrency can be located on the exterior of a triangle?

THINKING IT THROUGH

For each special segment, draw an example of an acute, an obtuse, and a right triangle. Construct the special segments for each classification of triangle to determine if their points of concurrency can be located on the exterior of a triangle.

Could a circumcenter be on the exterior of a triangle? _____

Could an incenter be on the exterior of a triangle? _____

Could a centroid be on the exterior of a triangle? _____

Could an orthocenter be on the exterior of a triangle? _____

The points of concurrency that can be on the exterior of a triangle are the _____ and the _____.

Lesson Practice

Choose the correct answer.

In Exercises 1–3 which point of concurrency is shown in the diagram?

1.

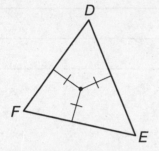

  (1) centroid
  (2) circumcenter
  (3) incenter
  (4) orthocenter

2.

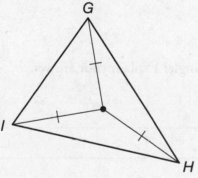

  (1) centroid
  (2) circumcenter
  (3) incenter
  (4) orthocenter

3. Find *JN*.

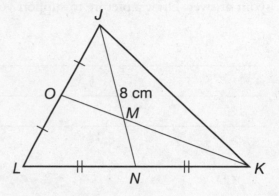

  (1) 16 cm
  (2) 12 cm
  (3) $\frac{51}{3}$ cm
  (4) 4 cm

4. The altitudes of a triangle are concurrent at a point called the

  (1) centroid.
  (2) circumcenter.
  (3) incenter.
  (4) orthocenter.

5. The medians of a triangle are concurrent at a point called the

  (1) centroid.
  (2) circumcenter.
  (3) incenter.
  (4) orthocenter.

OPEN-ENDED QUESTIONS

6. Which two types of special segments have the same point of concurrency in a right triangle? Explain your answer. Draw a picture to support your answer.

_____

_____

_____

7. Which point of concurrency is at a vertex of a right isosceles triangle? Explain your answer.

_____

_____

_____

_____

_____

23 Midsegment Theorem

G.G.42, G.G.43

A **midsegment** connects the midpoints of two sides of a triangle. In a triangle, the midsegment has special properties.

> **Midsegment Theorem:** The midsegment connecting two sides of a triangle is parallel to the third side and half its length.

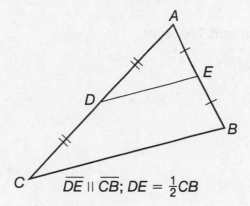

$\overline{DE} \parallel \overline{CB}$; $DE = \frac{1}{2}CB$

EXAMPLE 1

$\overline{OP}$, $\overline{PQ}$, and $\overline{QO}$ are midsegments. List pairs of parallel segments.

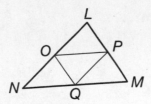

STRATEGY   Use the Midsegment Theorem.

$\overline{OP} \parallel \overline{MN}$

$\overline{QO} \parallel \overline{LM}$

$\overline{PQ} \parallel \overline{LN}$

SOLUTION   The pairs are listed above. The midsegments are parallel to the sides they do not touch.

EXAMPLE 2

Find the length of *FH* and *JK*.

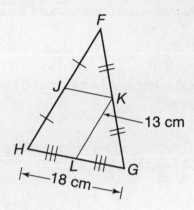

STRATEGY   Use the Midsegment Theorem.

STEP 1   Find the length of *FH*.

$$LK = \frac{1}{2}FH$$

$$13 = \frac{1}{2}FH$$

$$26 = FH$$

STEP 2   Find the length of *JK*.

$$JK = \frac{1}{2}GH$$

$$JK = \frac{1}{2}(18)$$

$$JK = 9$$

SOLUTION   *FH* = 26 cm, *JK* = 9 cm

EXAMPLE 3

Given: $\triangle LMN$ with midsegments $\overline{PQ}$, $\overline{QR}$, and $\overline{PR}$

Prove: $\triangle LPQ \cong \triangle PNR$

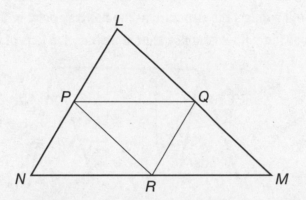

STRATEGY   Write a two-column proof.

Statements	Reasons
1. $\triangle LMN$ with midsegments $\overline{PQ}$, $\overline{QR}$, and $\overline{PR}$	1. Given
2. P is the midpoint of $\overline{LN}$.	2. Definition of midsegment
3. $\overline{LP} \cong \overline{PN}$	3. Definition of midpoint
4. $PQ = \frac{1}{2}NM$	4. Midsegment Theorem
5. R is the midpoint of $\overline{NM}$.	5. Definition of midsegment
6. $NR = \frac{1}{2}NM$	6. Definition of midpoint
7. $\frac{1}{2}NM = NR$	7. Symmetric Property
8. $PQ = NR$	8. Transitive Property of Equality
9. $\overline{PQ} \cong \overline{NR}$	9. Definition of congruent segments
10. $\overline{PQ} \parallel \overline{NM}$	10. Midsegment Theorem
11. $\angle LPQ \cong \angle PNR$	11. Corresponding Angles Theorem
12. $\triangle LPQ \cong \triangle PNR$	12. SAS Congruency Theorem

SOLUTION   The two-column proof is shown above.

COACHED EXAMPLE

Daren is setting up a swing set for his niece. For stability, he will bury the legs of the A-frame in concrete. The cross bars will support the structure at the halfway point in the angle. He bought crossbars that are 36 inches long. How far apart should he bury the legs of the A-frame?

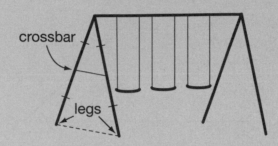

THINKING IT THROUGH

The crossbar represents the _____ of the triangle.

The midsegment is _____ the length of the base.

Write an equation and solve.

$$\underline{\hspace{4cm}} = \frac{1}{2} \left(\underline{\hspace{4cm}} \right)$$

Daren should bury the legs _____ inches apart.

Lesson Practice

Choose the correct answer.

Use the diagram for Exercises 1–3.

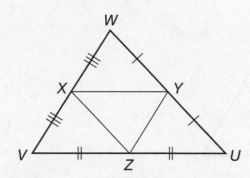

1. $\overline{WV}$ ‖ _?_

    (1) $\overline{WY}$

    (2) $\overline{YZ}$

    (3) $\overline{ZU}$

    (4) $\overline{XZ}$

2. $\overline{XZ}$ ‖ _?_

    (1) $\overline{WY}$

    (2) $\overline{YZ}$

    (3) $\overline{ZU}$

    (4) $\overline{XV}$

3. If $VZ = 20$ in., then $XY =$

    (1) 10 in.

    (2) 15 in.

    (3) 20 in.

    (4) 40 in.

4. Which angles in the diagram are congruent?

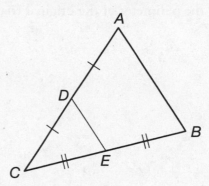

    (1) $\angle C \cong \angle A$

    (2) $\angle CDE \cong \angle CED$

    (3) $\angle CED \cong \angle EBA$

    (4) $\angle CBA \cong \angle CAB$

5. Given $HJ = 12$, $GF = 8$ and $HG = 7$, find the perimeter of $\triangle HJK$. (The interior segments are midsegments).

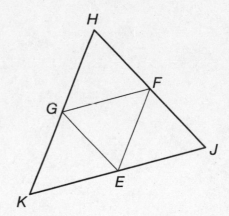

    (1) 21

    (2) 27

    (3) 35

    (4) 42

OPEN-ENDED QUESTION

6. Prove that the perimeter of a second triangle formed by the midsegments of an original triangle is half the perimeter of the original triangle.

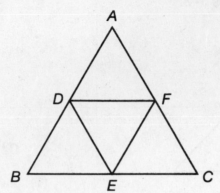

_____

_____

_____

_____

_____

Similar Triangles

G.G.44, G.G.45, G.G.46

Two polygons are **similar** if their sides are proportional. In the diagram, $\triangle ABC$ is similar to ($\sim$) $\triangle DEF$. Similar figures have congruent corresponding angles.

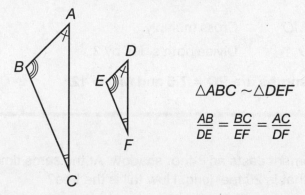

$\triangle ABC \sim \triangle DEF$

$$\frac{AB}{DE} = \frac{BC}{EF} = \frac{AC}{DF}$$

EXAMPLE 1

Given that $\triangle LMO \sim \triangle PQS$, find the unknown side lengths.

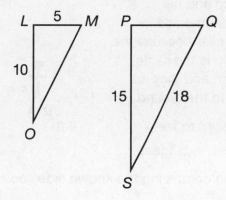

STRATEGY   Use proportions.

STEP 1   Set up the ratio of two corresponding side lengths that are known.

$\overline{LO}$ corresponds to $\overline{PS}$

$$\frac{LO}{PS} = \frac{10}{15} = \frac{2}{3}$$

STEP 2   Use the ratio to write a proportion involving $\overline{PQ}$.

$\overline{PQ}$ corresponds to $\overline{LM}$, so

$\frac{5}{PQ} = \frac{2}{3}.$

15 = 2PQ   Cross multiply.

7.5 = PQ   Divide both sides by 2.

STEP 3   Use the ratio to write a proportion involving $\overline{MO}$.

$\overline{QS}$ corresponds to $\overline{MO}$, so

$\frac{MO}{18} = \frac{2}{3}.$

36 = (3)MO   Cross multiply.

12 = MO   Divide both sides by 3.

SOLUTION   **The sides lengths are $PQ = 7.5$ and $MO = 12$.**

EXAMPLE 2

Erica is 5′6″ and at this moment casts an 8-foot shadow. At the same time, an old tree in her backyard casts a shadow that is 20 feet long. How tall is the tree?

STRATEGY   **Use properties of similar triangles.**

STEP 1   Draw a diagram.

Recognize that the triangles formed by Erica and her shadow and the tree and its shadow are similar because the angle of the sun is the same for both of them and they are perpendicular to the ground.

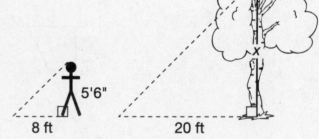

STEP 2   Convert Erica's height to feet.

5′6″ = 66 inches = 5.5 feet

STEP 3   Set up a proportion comparing the known sides and unknown length.

$\frac{8}{20} = \frac{5.5}{x}$

STEP 4   Solve the proportion.

110 = 8x   Cross multiply.

x = 13.75   Divide both sides by 8.

SOLUTION   **The tree is 13.75 feet tall.**

Although the definition of similar triangles requires that all three sides be proportional and all three angles be congruent, it is not necessary to know all six pieces of information to prove that triangles are similar.

Angle-Angle (AA) Similarity Postulate: If two angles of one triangle are congruent to two angles of another triangle, then the two triangles are similar.

Side-Side-Side (SSS) Similarity Theorem: If the lengths of three sides of one triangle are proportional to the lengths of three sides of another triangle, then the triangles are similar.

Side-Angle-Side (SAS) Similarity Theorem: If two sides of one triangle are proportional to two sides of another triangle, and the included angles are congruent then the triangles are similar.

EXAMPLE 3

Prove that the triangle formed by the midsegments of a triangle is similar to the original triangle.

STRATEGY   Write a two-column proof.

Given: U, V, and Z are the midpoints of $\overline{WX}$, $\overline{XY}$, and $\overline{WY}$ respectively.

Prove: $\triangle WXY \sim \triangle VZU$

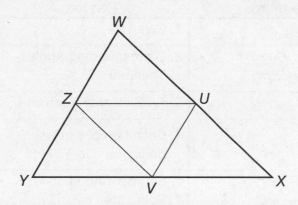

Statements	Reasons
1. U, V, and Z are the midpoints of $\overline{WX}$, $\overline{XY}$, and $\overline{WY}$ respectively.	1. Given
2. $ZU = \frac{1}{2}YX$	2. Midsegment Theorem
3. $ZV = \frac{1}{2}WX$	3. Midsegment Theorem
4. $UV = \frac{1}{2}WY$	4. Midsegment Theorem
5. $\triangle WXY \sim \triangle VZU$	5. SSS Similarity Theorem

SOLUTION   The two-column proof is shown above.

The Midsegment Theorem demonstrates a specific case of our next theorem.

> **Triangle Proportionality Theorem:** A segment is parallel to one side of a triangle if and only if it divides the other two sides proportionally.

EXAMPLE 4

Given: $\overline{PQ} \parallel \overline{CB}$

Prove: $\dfrac{AP}{PC} = \dfrac{AQ}{QB}$

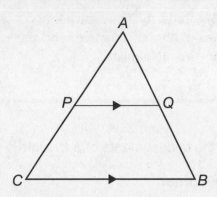

STRATEGY   Write a two-column proof.

Statements	Reasons
1. $\overline{PQ} \parallel \overline{CB}$	1. Given
2. $\angle APQ \cong \angle ACB$ $\angle AQP \cong \angle ABC$	2. Corresponding Angles Theorem
3. $\triangle APQ \sim \triangle ACB$	3. AA Similarity Postulate
4. $\dfrac{AP}{AC} = \dfrac{AQ}{AB}$	4. Definition of similar polygons
5. $AC = AP + PC$ $AB = AQ + QB$	5. Segment Addition Postulate
6. $\dfrac{AP}{AP + PC} = \dfrac{AQ}{AQ + QB}$	6. Substitution
7. $\dfrac{AP}{AP} + \dfrac{AP}{PC} = \dfrac{AQ}{AQ} + \dfrac{AQ}{QB}$	7. Distributive Property
8. $1 + \dfrac{AP}{PC} = 1 + \dfrac{AQ}{QB}$	8. Substitution
9. $\dfrac{AP}{PC} = \dfrac{AQ}{QB}$	9. Subtraction

SOLUTION   **The two-column proof is shown above.**

COACHED EXAMPLE

Given: $\overline{PQ} \parallel \overline{TS}$

Prove: $PR \cdot RT = QR \cdot RS$

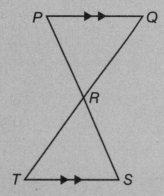

THINKING IT THROUGH

What is always the first statement in a two-column proof?_____

What types of angles are $\angle PRQ$ and $\angle SRT$?_____

What angle is alternate interior with $\angle RPQ$?_____

How do you know that the sides of similar triangles are proportional?_____

The final statement of a proof is the _____.

Statements	Reasons
1._____	1. Given
2. $\angle PRQ \cong \angle SRT$	2._____
3. $\angle RPQ \cong$ _____	3. Alternate Interior Angles
4._____	4. AA Similarity Postulate
5. $\dfrac{RP}{RS} = \dfrac{RQ}{RT}$	5._____
6._____	6. Multiplication Property

Complete the proof above.

Lesson Practice

Choose the correct answer.

1. Which types of triangles are always similar?

- **(1)** isosceles
- **(2)** right
- **(3)** equilateral
- **(4)** acute

2. Which of the following statements is always true about similar triangles?

- **(1)** Similar triangles have parallel bases.
- **(2)** Similar triangles are the same shape.
- **(3)** Similar triangles are the same size.
- **(4)** Similar triangles have congruent sides.

3. In the diagram, $\overline{HE} \parallel \overline{GF}$. Find DF.

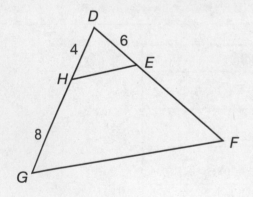

- **(1)** 3
- **(2)** 12
- **(3)** 18
- **(4)** 24

Use the diagram for Exercises 4–6.

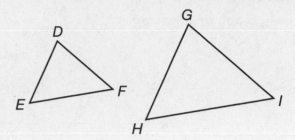

4. $\triangle DEF \sim \triangle GHI$. $DE = 4$, $DF = 6$, and $GH = 10$. Find GI.

- **(1)** 2.4
- **(2)** 6.7
- **(3)** 12
- **(4)** 15

5. Given $\dfrac{DE}{GH} = \dfrac{EF}{HI}$, what other piece of information would you need to prove $\triangle DEF \sim \triangle GHI$?

- **(1)** $\angle F \cong \angle I$
- **(2)** $\angle E \cong \angle H$
- **(3)** $\overline{DE} \cong \overline{GH}$
- **(4)** $\overline{EF} \cong \overline{HI}$

6. Given $\angle D \cong \angle G$, what other information would not help prove $\triangle DEF \sim \triangle GHI$?

- **(1)** $\angle E \cong \angle H$
- **(2)** $\angle F \cong \angle I$
- **(3)** $\dfrac{DE}{GH} = \dfrac{DF}{GI}$
- **(4)** $\dfrac{DE}{DF} = \dfrac{DE}{EF}$

OPEN-ENDED QUESTION

7. Prove that if the vertex angle of an isosceles triangle is congruent to the vertex angle of another isosceles triangle, then the triangles are similar.

_____

_____

_____

_____

_____

25 Right Triangles

G.G.44, G.G.45, G.G.46, G.G.47

A **right triangle** is a triangle with one right angle. The side opposite the right angle is the **hypotenuse**. The other two sides are **legs**.

The **mean proportional** of two numbers is the number x such that $\frac{a}{x} = \frac{x}{b}$ or $x = \sqrt{a \cdot b}$. For example, the mean proportional of 16 and 4 is 8 because $16 \cdot 4 = 64$ and $\sqrt{64} = 8$.

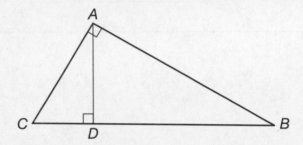

Right Triangle-Altitude Theorem: If the altitude of a right triangle is drawn, then it divides the triangle into two similar triangles that are also similar to the original.
In the diagram, $\triangle ABC \sim \triangle DBA \sim \triangle DAC$.

Mean Proportionality Theorem: The altitude of a right triangle is the mean proportional between the segments into which it divides the hypotenuse.
In the diagram, $\frac{CD}{AD} = \frac{AD}{DB}$.

Leg Proportionality Theorem: Each leg of a right triangle is the mean proportional between the hypotenuse and the corresponding segment of the hypotenuse.
In the diagram, $\frac{CB}{AC} = \frac{AC}{CD}$ and $\frac{CB}{AB} = \frac{AB}{DB}$.

EXAMPLE 1

Prove the Right Triangle-Altitude Theorem.

STRATEGY   **Write a two-column proof.**

Given: $\triangle DEF$ is a right triangle. $\overline{DG}$ is the altitude of $\triangle DEF$.

Prove: $\triangle DEF \sim \triangle GED$; $\triangle GDF \sim \triangle DEF$; $\triangle GED \sim \triangle GDF$

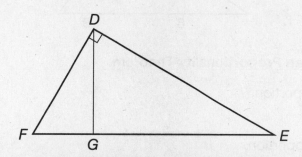

Statements	Reasons
1. $\triangle DEF$ is a right triangle. $\overline{DG}$ is the altitude of $\triangle DEF$.	1. Given
2. $\angle DGE$ is a right angle.	2. Definition of altitude
3. $\triangle GED$ is a right triangle.	3. Definition of right triangle
4. $\angle FDE \cong \angle DGE$	4. All right angles are congruent.
5. $\angle E \cong \angle E$	5. Reflexive Property
6. $\triangle DEF \sim \triangle GED$	6. AA Similarity Postulate
7. $\angle DGF$ is a right triangle.	7. Definition of altitude
8. $\angle FDE \cong \angle DGF$	8. All right angles are congruent.
9. $\angle F \cong \angle F$	9. Reflexive Property
10. $\triangle GDF \sim \triangle DEF$	10. AA Similarity Postulate
11. $\angle F$ is complementary to $\angle GDF$.	11. Definition of complementary angles
12. $\angle GDE$ is complementary to $\angle GDF$.	12. Definition of complementary angles
13. $\angle F \cong \angle GDE$	13. Substitution Theorem
14. $\angle DGE \cong \angle DGF$	14. All right angles are congruent.
15. $\triangle GED \sim \triangle GDF$	15. AA Similarity Postulate

SOLUTION   **The two-column proof is shown above.**

EXAMPLE 2

Find the value of x.

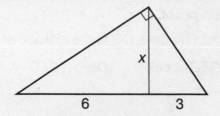

STRATEGY   Use the Mean Proportionality Theorem.

STEP 1   Set up a proportion.

$$\frac{6}{x} = \frac{x}{3}$$

STEP 2   Solve the proportion.

$$x^2 = 18$$
$$\sqrt{x^2} = \sqrt{18}$$
$$x = 3\sqrt{2}$$

SOLUTION   The length of the altitude is $3\sqrt{2}$ or approximately 4.2.

EXAMPLE 3

Find *IJ* and *HG*.

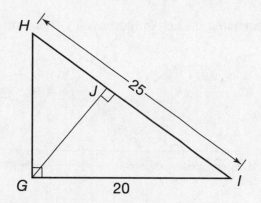

STRATEGY Use the Leg Proportionality Theorem.

STEP 1 Set up a proportion to find *IJ*.

$$\frac{HI}{GI} = \frac{GI}{JI}$$

$$\frac{25}{20} = \frac{20}{JI}$$

STEP 2 Solve the proportion.

$$(25)JI = 400$$

$$JI = 16$$

STEP 3 Find *HJ*.

$$HJ = HI - JI$$

$$HJ = 25 - 16 = 9$$

STEP 4 Set up a proportion to find *HG*.

$$\frac{HI}{HG} = \frac{HG}{HJ}$$

$$\frac{25}{HG} = \frac{HG}{9}$$

STEP 5 Solve the proportion.

$$HG^2 = 225$$

$$\sqrt{HG^2} = \sqrt{225}$$

$$HG = 15$$

SOLUTION **JI = 16, and HG = 15**

COACHED EXAMPLE

Prove the Pythagorean Theorem using the Leg Proportionality Theorem.

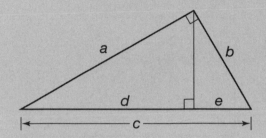

THINKING IT THROUGH

Write proportions using the Leg Proportionality Theorem.

_____   _____

Cross multiply to get expressions for a^2 and b^2.

_____   _____

Use your expressions to complete the algebraic proof.

$a^2 + b^2 =$ _____

$a^2 + b^2 =$ _____   Factor out c.

$a^2 + b^2 =$ _____   What is $d + e$ equal to?

$a^2 + b^2 = c^2$ The Pythagorean Theorem

The proof is complete.

Lesson Practice

Choose the correct answer.

In Exercises 1–4, find the value of each variable. If necessary, round to the nearest tenth.

1.

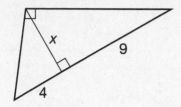

(1) 1.8

(2) 2.3

(3) 6

(4) 20.3

2.

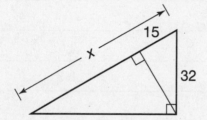

(1) 2.1

(2) 7

(3) 21.9

(4) 68.3

3.

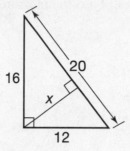

(1) 7.2

(2) 9.6

(3) 12.8

(4) 18

4.

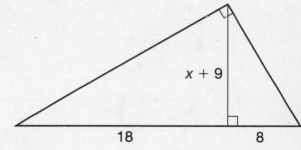

(1) 3

(2) 7

(3) 9

(4) 12

5. In a right triangle, the measure of the hypotenuse is 10 inches. The measure of the shorter leg is 4 inches. Find the measure of the longer leg.

    (1) 1.6 inches

    (2) 8.4 inches

    (3) 9.2 inches

    (4) 12 inches

6. In a right triangle, the altitude is 5 units and the longest segment of the hypotenuse is 7 units. Find the length of the hypotenuse.

    (1) 3.6

    (2) 5.9

    (3) 9.8

    (4) 10.6

OPEN-ENDED QUESTION

7. The altitude in a right triangle is 6 cm. The altitude divides the hypotenuse into segments whose lengths are in the ratio 1:4. Find the length of the two segments.

_____

_____

_____

_____

_____

26 The Pythagorean Theorem

G.G.48

The **Pythagorean Theorem** states that in a right triangle, the sum of the squares of the lengths of the legs is equal to the square of the lengths of the hypotenuse. A **Pythagorean triple** is a set of three whole numbers that satisfy the Pythagorean Theorem and could represent the lengths of the sides of a right triangle. Some common Pythagorean triples are (3, 4, 5), (5, 12, 13), (7, 24, 25), (8, 15, 17), and (9, 40, 41).

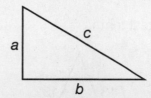

Pythagorean Theorem
$a^2 + b^2 = c^2$

EXAMPLE 1

Find the value of x.

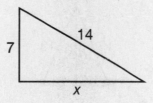

STRATEGY   Use the Pythagorean Theorem.

   STEP 1   Set up an equation.

              The legs are labeled 7 and x.

              The hypotenuse is 14.

              $7^2 + x^2 = 14^2$

STEP 2   Solve the equation for x.

$$49 + x^2 = 196$$ Simplify.

$$x^2 = 147$$ Subtract.

$$\sqrt{x^2} = \sqrt{147}$$ Take the square root of both sides to solve for x.

$$x = \sqrt{147}$$ Find a perfect square factor of 147.

$$x = \sqrt{49} \cdot \sqrt{3}$$

$$x = 7\sqrt{3}$$

SOLUTION   **The length of the other leg is $7\sqrt{3}$ or approximately 12.1.**

EXAMPLE 2

Find the area of the triangle to the nearest tenth.

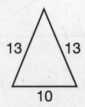

STRATEGY   **Use the Pythagorean Theorem.**

STEP 1   Find the height of the triangle.

You are not given the height, but since the triangle is isosceles, the altitude divides it into two congruent right triangles with a leg of 5.

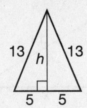

This is a Pythagorean triple 5, 12, 13. The height is 12.

STEP 2   Calculate the area of the triangle.

$$A = \tfrac{1}{2}bh = \tfrac{1}{2}(10)(12) = 60$$

SOLUTION   **The area of the triangle is 60 square units.**

The converse of the Pythagorean Theorem can also be used to prove that a triangle is a right triangle.

> **Pythagorean Theorem Converse:** In a triangle, if the sum of the squares of the lengths of two sides is equal to the square of the length of the third side, then it is a right triangle.

EXAMPLE 3

Show that the triangle is a right triangle.

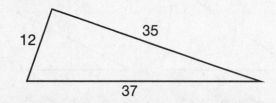

STRATEGY   **Use the converse of the Pythagorean Theorem.**

STEP 1   Decide which sides should be the legs.

The legs of a right triangle are shorter than the hypotenuse, so the sides marked 12 and 35 should be the legs.

STEP 2   Determine if the Pythagorean Theorem is true for this triangle.

$$12^2 + 35^2 \stackrel{?}{=} 37^2$$
$$144 + 1225 \stackrel{?}{=} 1369$$
$$1369 = 1369 \checkmark$$

SOLUTION   **The triangle is a right triangle because it satisfies the Pythagorean Theorem.**

The Pythagorean Theorem even has applications with triangles other than right triangles.

Theorem 26.1: In a triangle, if the square of the length of the longest side is less than the sum of the squares of the lengths of the other two sides, then it is acute.

Theorem 26.2: In a triangle, if the square of the lengths of the longest side is greater than the sum of the squares of the lengths of the other two sides, then it is obtuse.

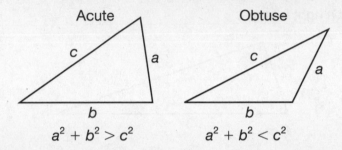

Acute
$$a^2 + b^2 > c^2$$

Obtuse
$$a^2 + b^2 < c^2$$

EXAMPLE 4

Do the side lengths 10, 11, and 14 form a triangle? If so, what type of triangle?

STRATEGY   Use the Triangle Inequality and Theorems 26.1 and 26.2.

STEP 1   Determine if the lengths can form a triangle.

If the lengths can form a triangle, then they will satisfy the triangle inequality. Check that each side is greater than the difference and less than the sum of the other two sides.

$$11 - 10 < 14 < 10 + 11$$
$$14 - 11 < 10 < 11 + 14$$
$$14 - 10 < 11 < 14 + 10$$

The lengths can form a triangle.

STEP 2   Determine if the square of the largest side is greater or less than the sum of the squares of the other two sides.

$$14^2 = 196$$
$$10^2 + 11^2 = 100 + 121 = 221 > 196$$

SOLUTION   Because the square of the longest side is less than the sum of the squares of the other two, the triangle is acute.

COACHED EXAMPLE

Khalil regularly takes a shortcut across the park instead of walking along the sidewalk on his way home from school. How much distance does he cut off of his trip by taking the shortcut?

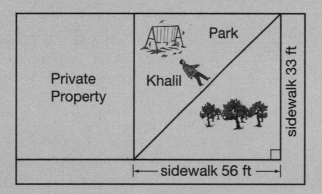

THINKING IT THROUGH

How far would Khalil have to walk if he stayed on the sidewalk? _____

The path Khalil takes across the park is the _____ of a right triangle.

Use the _____ Theorem to find the length of the _____.

$c^2 =$ _____$^2 +$ _____^{2}

$c^2 =$ _____ $+$ _____

$c^2 =$ _____

$c =$ _____

Find the difference in the length of the sidewalk path and the shortcut.

_____

Khalil walks _____ less feet on his shortcut.

Lesson Practice

Choose the correct answer.

1. The legs of a right triangle are 3 and 4 inches long. How long is the hypotenuse?

   (1) 5 inches

   (2) 7 inches

   (3) 13 inches

   (4) 25 inches

2. The hypotenuse of a right triangle is 89 units. One leg is 39 units. Find the length of the other leg.

   (1) 50 units

   (2) 62 units

   (3) 80 units

   (4) 97 units

3. A cat is stuck 15 feet up in a tree. To get it down, you will place a ladder 5 feet from the base of the tree. How tall must your ladder be in order to reach the cat? Round to the nearest foot.

   (1) 10 ft

   (2) 14 ft

   (3) 15 ft

   (4) 16 ft

4. A baseball field is a square with side length of 90 ft. What is the shortest distance from first to third base? Round to the nearest tenth.

   (1) 9.5 ft

   (2) 13.4 ft

   (3) 90 ft

   (4) 127.3 ft

5. A square field has a 25-foot sprinkler system running the length of the diagonal. What is the area of the field?

   (1) 17.7 ft^2

   (2) 25 ft^2

   (3) 312.5 ft^2

   (4) 625 ft^2

For Exercises 6 and 7 decide whether the numbers can represent the side lengths of a triangle. If so, what type of triangle?

6. 10, 49, 50

    (1) not a triangle

    (2) acute

    (3) right

    (4) obtuse

7. 2, 10, 12

    (1) not a triangle

    (2) acute

    (3) right

    (4) obtuse

OPEN-ENDED QUESTION

8. An equilateral triangle has a side length of 7 meters. Find the area of the triangle to the nearest tenth.

_____

_____

_____

_____

_____

Polygons

G.G.36, G.G.37

A **polygon** is a closed plane figure made up of line segments that meet at their endpoints with at least three sides. A **convex polygon** is a polygon in which no diagonals contain points outside the polygon. A **concave polygon** has diagonals outside the polygon. **Interior angles** are on the inside of the polygon at each vertex. **Exterior angles** are formed by extending one side outside the polygon at each vertex. The interior and exterior angles at a vertex are supplementary. A **regular polygon** is equilateral (all sides congruent) and equiangular (all angles congruent).

Polygons	Not Polygons
Interior Angle Convex Polygon	
Exterior Angle Concave Polygon	
Regular Polygon	

> **Interior Angle Sum Theorem:** The sum of the measures of interior angles of a polygon is
>
> $S_n = (n - 2)180°$ where n is the number of sides of the polygon.

EXAMPLE 1

Use the formula to determine the sum of the interior angles in a rectangle.

STRATEGY **Use the Interior Angle Sum Theorem.**

  STEP 1 Determine the value of n.

  A rectangle is a quadrilateral.

  Quadrilaterals have four sides.

  $n = 4$

  STEP 2 Substitute the value into the formula.

  $(4 - 2)180° = 2(180°) = 360°$

SOLUTION **The sum of the interior angles in a quadrilateral is 360°.**

EXAMPLE 2

What is the measure of one interior angle in a regular hexagon?

STRATEGY **Use the Interior Angles Sum Theorem**

  STEP 1 Determine the value of n.

  A hexagon has 6 sides. $n = 6$

  STEP 2 Substitute the value for n into the formula.

  $(6 - 2)180° = 4(180) = 720°$

  STEP 3 Because all interior angles are congruent in a regular polygon, divide the sum by 6 to find the measure of one angle.

  $\frac{720°}{6} = 120°$

SOLUTION **One angle in a regular hexagon has a measure of 120°.**

Exterior Angle Sum Theorem: The sum of the measures of the exterior angles of any polygon is 360°.

EXAMPLE 3

Find the measure of an exterior angle of a regular pentagon.

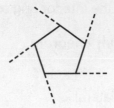

STRATEGY   **Use the Exterior Angle Sum Theorem.**

The sum of the exterior angles in a polygon is always 360°. In a pentagon, there are five exterior angles.

$$\frac{360°}{5} = 72°$$

SOLUTION   **One exterior angle in a regular pentagon measures 72°.**

The **apothem** is the height of the triangle formed by connecting the center of the polygon to each vertex of a pair of adjacent vertices. In a regular polygon the apothem bisects the central angle of the triangle. The area of a regular polygon is found by taking one-half the product of the length of the apothem and the perimeter. To calculate the perimeter, multiply the side length by the number of sides.

Area of a Regular Polygon

$A = \frac{1}{2}aP$ where a = length of the apothem, P = perimeter

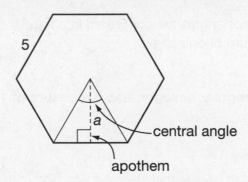

In the study of polygons, a **central angle** is an angle whose vertex is the center of a polygon and whose sides connect the center to each vertex of a pair of adjacent vertices.

EXAMPLE 4

Find the area of a regular pentagon with a side length of 6 cm.

STRATEGY   **Use trigonometry to find the length of the apothem.**

STEP 1   Find the measure of a central angle in the pentagon.

There are 5 central angles around the 360° center point.

$$\frac{360°}{5} = 72°$$

STEP 2   Find the measure of the vertex angle.

The apothem bisects the central angle and the side of the pentagon. In this problem, it forms one leg of a right triangle with one 36° angle and opposite side length of 3 cm.

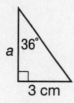

STEP 3   Use trigonometry to find the length of the apothem.

$\tan 36° = \frac{3}{a}$ Use the tangent ratio.

$a \cdot \tan 36° = 3$ Multiply by a.

$a = \frac{3}{\tan 36°}$ Divide by $\tan 36°$ to isolate a.

$a \approx 4.13$ cm   Simplify.

STEP 4   Use the formula for the area of a regular polygon.

$A = \frac{1}{2}(4.13)(5 \cdot 6) = 61.95$

SOLUTION   **The area of the pentagon is approximately 61.95 cm^2.**

COACHED EXAMPLE

What is the sum of the interior angles of an octagon?

THINKING IT THROUGH

Write the formula for the sum of the interior angles of a polygon. _____

Write the value of n, the number of sides for an octagon _____.

Write the value of $n - 2$. _____

Write the product of $(n - 2)180°$. _____

The sum of the interior angles of an octagon is _____.

Lesson Practice

Choose the correct answer.

1. All of the following polygons have an interior angle sum of 180° except

   (1) equilateral triangle.

   (2) scalene triangle.

   (3) obtuse triangle.

   (4) trapezoid.

2. Find the area of a regular hexagon with side length of 10 in.

   (1) 101.3 in.^{2}

   (2) 150 in.^{2}

   (3) 259.8 in.^{2}

   (4) 519.6 in.^{2}

3. What is the sum of the interior angles of a convex quadrilateral?

   (1) 180°

   (2) 270°

   (3) 360°

   (4) 540°

4. A decagon has ten sides. How many degrees are in its interior angles?

   (1) 1,000°

   (2) 1,080°

   (3) 1,440°

   (4) 1,800°

5. When does the measure of the interior angle of a polygon and its adjacent exterior angle equal 180°?

   (1) always

   (2) sometimes

   (3) never

   (4) cannot be determined

6. If the sum of two exterior angles of a triangle equal 249 degrees, then what is the measure of the third exterior angle?

   (1) 111°

   (2) 180°

   (3) 249°

   (4) 360°

7. Find the measure of one exterior angle of a regular octagon.

   (1) 30°

   (2) 45°

   (3) 60°

   (4) 135°

8. A regular octagon is inscribed in a circle with a radius of 5 feet. What is the area of the octagon?

   (1) 49.68 ft^2

   (2) 70.67 ft^2

   (3) 110.88 ft^2

   (4) 120 ft^2

OPEN-ENDED QUESTIONS

9. If one interior angle of a pentagon is 90 degrees, then what is the sum of the remaining interior angles? Explain your answer.

_____

_____

_____

10. Can the Interior Angle Sum Theorem be used for a concave polygon?

_____

Explain your answer.

_____

_____

_____

_____

_____

Parallelograms

A **parallelogram** is a quadrilateral with opposite sides parallel. In a quadrilateral, you can draw a **diagonal**, or a segment connecting two nonconsecutive vertices.

The table outlines several ways to prove a quadrilateral is a parallelogram.

Theorem	Conditions
28.1	A quadrilateral is a parallelogram if and only if the diagonals bisect each other.
28.2	A quadrilateral is a parallelogram if and only if one pair of opposite sides is both congruent and parallel.
28.3	A quadrilateral is a parallelogram if and only if both pairs of opposite angles are congruent.
28.4	A quadrilateral is a parallelogram if and only if both pairs of opposite sides are congruent.
28.5	A quadrilateral is a parallelogram if and only if each angle is supplementary to both its consecutive angles.

EXAMPLE 1

Prove that if the diagonals of a quadrilateral bisect each other, then it is a parallelogram.

Given: $\overline{AC}$ and $\overline{BD}$ bisect each other at E.

Prove: $ABCD$ is a parallelogram.

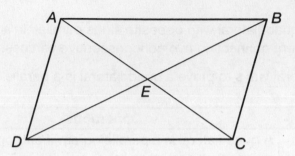

STRATEGY   Write a two-column proof.

Statements	Reasons
1. $\overline{AC}$ and $\overline{BD}$ bisect each other at E.	1. Given
2. $\overline{AE} \cong \overline{CE}$ $\overline{BE} \cong \overline{DE}$	2. Definition of segment bisector
3. $\angle AEB \cong \angle CED$ $\angle BEC \cong \angle DEA$	3. Vertical Angles Theorem
4. $\triangle AEB \cong \triangle CED$ $\triangle BEC \cong \triangle DEA$	4. SAS Congruence Theorem
5. $\angle BAE \cong \angle DCE$ $\angle ECB \cong \angle EAD$	5. CPCTC
6. $\overline{AB} \parallel \overline{CD}$ $\overline{BC} \parallel \overline{AD}$	6. Converse of Alternate Interior Angles Theorem
7. $ABCD$ is a parallelogram.	7. Definition of parallelogram

SOLUTION   **The two-column proof is shown above.**

EXAMPLE 2

Prove that if both pairs of opposite angles in a quadrilateral are congruent, then it is a parallelogram.

Given: $\angle E \cong \angle G$ and $\angle F \cong \angle H$

Prove: *EFGH* is a parallelogram.

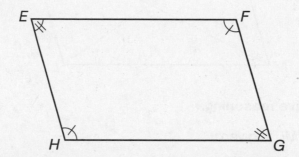

STRATEGY   Write a two-column proof.

Statements	Reasons
1. $\angle E \cong \angle G$ and $\angle F \cong \angle H$	1. Given
2. $(2)(\angle E) + (2)(\angle F) = 360$	2. The sum of the interior angles of a quadrilateral is 360°.
3. $(2)(\angle F) = 360 - 2(\angle E)$	3. Subtraction Property
4. $\angle F = 180 - \angle E$	4. Division Property
5. $\angle E + \angle F = 180°$ $\angle G + \angle H = 180$	5. Definition of supplementary angles
6. $\overline{EF} \parallel \overline{HG}$ $\overline{FG} \parallel \overline{EH}$	6. Converse of Consecutive Interior Angles Theorem
7. *ABCD* is a parallelogram.	7. Definition of a parallelogram

SOLUTION   The two-column proof is shown above.

EXAMPLE 3

Use the diagram to prove that quadrilateral *JKLM* is a parallelogram given △*PKJ* ≅ △*PML*.

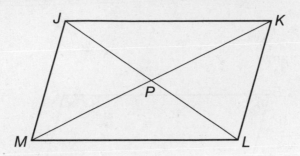

STRATEGY   **Use deductive reasoning.**

STEP 1   △*PKJ* ≅ △*PML* is given.

STEP 2 $\overline{PK} \cong \overline{PM}$, and $\overline{PJ} \cong \overline{PL}$ by CPCTC.

STEP 3 $\overline{JL}$ and $\overline{KM}$ bisect each other because of the definition of segment bisector.

SOLUTION   ***JKLM* is a parallelogram.**

To find the area of a parallelogram, use its height. The height is a segment drawn from one vertex perpendicular to the other side.

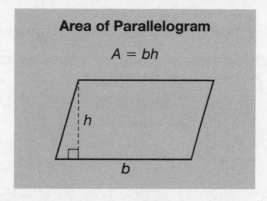

Area of Parallelogram

$$A = bh$$

EXAMPLE 4

Find the area of a parallelogram with a base of 8 cm and a height of 3 cm.

STRATEGY   **Use the formula for the area of a parallelogram.**

$A = bh$

$A = 8 \cdot 3$

$A = 24$

SOLUTION   **The area of the parallelogram is 24 cm^2.**

COACHED EXAMPLE

Complete the proof by providing the correct missing statement or reason.

Given: Parallelogram *ABCD*; *E* is the midpoint of $\overline{BC}$ and *F* is the midpoint of $\overline{AD}$.

Prove: *ABEF* is a parallelogram.

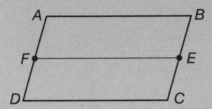

THINKING IT THROUGH

The reason in the first statement of a proof is usually _____.

Opposite sides of a parallelogram are parallel and _____.

To replace an expression with one that is equal to it is permitted by the
_____ Property.

If $ac = bc$, then $a = b$ is which property of equality? _____

If one pair of the opposite sides of a quadrilateral is both congruent and parallel, then the quadrilateral
is a _____.

Statements	Reasons
1. *E* is the midpoint of $\overline{BC}$ and F is the midpoint of $\overline{AD}$. ABCD is a parallelogram.	1. _____
2. $\overline{AD} \cong \overline{BC}$	2. _____
3. $\overline{AD} = 2\overline{AF}$ and $\overline{BC} = 2\overline{BE}$	3. Definition of midpoint
4. _____	4. Substitution Property
5. $\overline{AF} = \overline{BE}$	5. _____
6. _____	6. A quadrilateral is a parallelogram if and only if one pair of opposite sides is both congruent and parallel.

Lesson Practice

Choose the correct answer.

1. Find the area of the parallelogram.

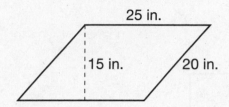

25 in.

15 in.   20 in.

   (1) 200 in.^{2}

   (2) 300 in.^{2}

   (3) 375 in.^{2}

   (4) 500 in.^{2}

2. The length of one side of a parallelogram is 14 inches. The length of the opposite side must be

   (1) 11 inches.

   (2) 12 inches.

   (3) 13 inches.

   (4) 14 inches.

3. In parallelogram *ABCD*, the ratio of the measures of angles *A* and *B* is 3:7. Find the measure of $\angle A$.

   (1) 26°

   (2) 54°

   (3) 60°

   (4) 126°

4. In parallelogram *PQRS*, $\angle PQR = 76°$. If $\angle RSP$ is opposite $\angle PQR$, then $\angle RSP =$

   (1) 76°.

   (2) 104°.

   (3) 180°.

   (4) 360°.

5. Quadrilateral *UVWX* has two intersecting diagonals. If the quadrilateral is a parallelogram, the diagonals must

   (1) be perpendicular.

   (2) be parallel.

   (3) bisect each other.

   (4) coincide.

6. A quadrilateral has one pair of opposite sides parallel. What additional information would prove that the quadrilateral is a parallelogram?

   (1) Diagonals are congruent.

   (2) The parallel sides are also congruent.

   (3) The other pair of sides are congruent.

   (4) Two angles are supplementary.

OPEN-ENDED QUESTION

7. Write a two-column proof to show that $RSTW$ is a parallelogram given $\triangle TRS \cong \triangle RTW$ with diagonal $\overline{RT}$.

    Given: $\triangle TRS \cong \triangle RTW$

    Prove: $RSTW$ is a parallelogram.

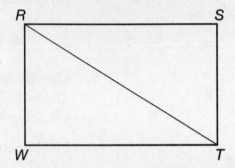

29 Special Parallelograms

Some parallelograms have special characteristics. A **rectangle** is a parallelogram with opposite sides congruent and four right angles. A **rhombus** has four congruent sides. A **square** has four congruent sides and four right angles.

> **Theorem 29.1**: A parallelogram is a rectangle if and only if it contains four right angles.
>
> **Theorem 29.2**: A parallelogram is a rectangle if and only if its diagonals are congruent.

EXAMPLE 1

Prove that if a figure is a rectangle, then its diagonals are congruent.

STRATEGY   Write a two-column proof.

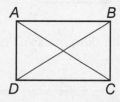

Given: *ABCD* is a rectangle

Prove: $\overline{AC} \cong \overline{BD}$

Statements	Reasons
1. *ABCD* is a rectangle	1. Given
2. $\overline{AB} \cong \overline{DC}$ and $\overline{AD} \cong \overline{BC}$	2. Opposite sides of a rectangle are congruent.
3. ∠*ABC* and ∠*DCB* are right angles.	3. A rectangle has all right angles.
4. ∠*ABC* ≅ ∠*DCB*	4. All right angles are congruent.
5. △*ABC* ≅ △*DCB*	5. SAS Congruence Theorem
6. $\overline{AC} \cong \overline{DB}$	6. CPCTC

SOLUTION   The two-column proof is shown above.

Theorem 29.3: A parallelogram is a rhombus if and only if two consecutive sides are congruent.

Theorem 29.4: A parallelogram is a rhombus if and only if the diagonals bisect the angles.

Theorem 29.5: A parallelogram is a rhombus if and only if the diagonals are perpendicular.

Theorem 29.6: A parallelogram is a square if and only if it is a rectangle and a rhombus.

EXAMPLE 2

Prove that if a figure is a rhombus, then the diagonals are perpendicular.

STRATEGY   Write a two-column proof.

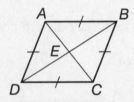

Given: Rhombus *ABCD* with diagonals *AC* and *BD*

Prove: $\overline{AC} \perp \overline{BD}$

Statements	Reasons
1. *ABCD* is a rhombus.	1. Given
2. $\overline{AB} \cong \overline{BC} \cong \overline{CD} \cong \overline{DA}$	2. A rhombus is equilateral.
3. $\overline{DE} \cong \overline{BE}$	3. A rhombus is a parallelogram, therefore the diagonals bisect each other.
4. $\overline{AE} \cong \overline{AE}$	4. Reflexive Property
5. $\triangle ADE \cong \triangle ABE$	5. SSS Congruence Theorem
6. $\angle AED \cong \angle AEB$	6. CPCTC
7. $\overline{AC} \perp \overline{BD}$	7. When two lines meet to form congruent, adjacent angles, the lines are perpendicular.

SOLUTION   The two-column proof is shown above.

EXAMPLE 3

Find the value of x that would make the parallelogram a rhombus.

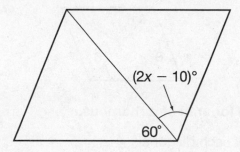

STRATEGY   Use Theorem 29.4.

STEP 1   Set up an equation and solve for x.
The diagonals of a rhombus bisect the angles. The two values are equal.

$$60 = 2x - 10$$
$$70 = 2x$$
$$x = 35$$

STEP 3   Check your answer by substituting the value into the expression.

$$2(35) - 10$$
$$70 - 10$$
$$60$$

The measures of the angles are equal.

SOLUTION   **The value that makes the parallelogram a rhombus is $x = 35$.**

To find the area of a rectangle or square, simply multiply the side lengths.

> **Area of a Rectangle**
>
> $A = lw$

To find the area of a rhombus you can use the formula for the area of a parallelogram. Or, use a formula that is specific to rhombi that uses the length of the diagonals.

> **Area of a Rhombus**
>
> $A = \frac{1}{2}d_1 d_2$

EXAMPLE 4

Find the area of the rhombus.

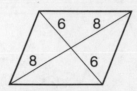

STRATEGY   Use the formula for area of a rhombus.

    STEP 1   Find the length of each diagonal.

$$6 + 6 = 12$$
$$8 + 8 = 16$$

    STEP 2   Substitute the values into the formula.

$$A = \frac{1}{2}(12)(16) = 96$$

SOLUTION   The area of the rhombus is 96 square units.

COACHED EXAMPLE

Complete the table to identify the characteristics of special parallelograms.

THINKING IT THROUGH

Place an X to indicate each figure that has that characteristic.

Property	Rectangle	Rhombus	Square
Opposite sides are parallel.			
Diagonals are perpendicular.			
Diagonals are congruent.			
All sides are congruent.			
Diagonals bisect each other.			
All angles are congruent.			

Lesson Practice

Choose the correct answer.

1. If the diagonals of a parallelogram are congruent, then the parallelogram is a

   (1) polygon.

   (2) rectangle.

   (3) rhombus.

   (4) trapezoid.

2. In parallelogram *PQRS*, diagonals *PR* and *QS* are perpendicular at point *T*. If *PT* = 8 and *TS* = 15, find *QR*.

   (1) 8

   (2) 15

   (3) 17

   (4) 20

3. In parallelogram *RSTW*, $\overline{RT}$ and $\overline{SW}$ bisect each other at *A* and $\overline{RA} \cong \overline{WA}$. *RSTW* must be a

   (1) parallelogram.

   (2) rectangle.

   (3) rhombus.

   (4) trapezoid.

4. Which of the following groups is a square not a member of?

   (1) parallelograms

   (2) rectangles

   (3) rhombi

   (4) trapezoids

5. The diagonals of a parallelogram are congruent and they bisect the angles. Which type of special parallelogram is it?

   (1) rectangle

   (2) rhombus

   (3) hexagon

   (4) trapezoid

6. What is the area of a rhombus with diagonals of 8 inches and 12 inches?

   (1) 20 in.^{2}

   (2) 32 in.^{2}

   (3) 48 in.^{2}

   (4) 96 in.^{2}

7. A square has a perimeter of 60 mm. What is the area of the square?

(1) 30 mm^2

(2) 60 mm^2

(3) 225 mm^2

(4) 900 mm^2

8. The diagonals of a rhombus are 6 in. and 8 in. Find the perimeter of the rhombus.

(1) 28 in.

(2) 20 in.

(3) 16 in.

(4) 12 in.

OPEN-ENDED QUESTION

9. The figure is a rectangle. Find the measure of $\angle 2$.

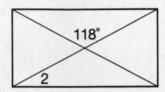

_____

Explain your reasoning.

_____

_____

_____

_____

30 Trapezoids

A **trapezoid** is a quadrilateral with exactly one pair of parallel sides, or **bases**. The other two sides are **legs**. Because of the parallel sides, a trapezoid has two pairs of consecutive angles that are supplementary.

An **isosceles trapezoid** has two legs that are congruent and base angles that are congruent.

A **right trapezoid** has exactly two right angles.

Trapezoids

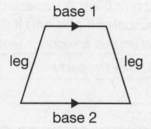

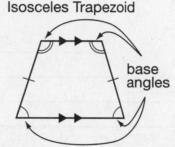

Isosceles Trapezoid

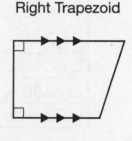

Right Trapezoid

Isosceles trapezoids have several unique characteristics.

Theorem 30.1: A trapezoid is isosceles if and only if the base angles are congruent.

Theorem 30.2: A trapezoid is isosceles if and only if the diagonals are congruent.

Theorem 30.3: If a trapezoid is isosceles, then the opposite angles are supplementary.

EXAMPLE 1

Given: *ABCD* is an isosceles trapezoid and $\overline{AB} \cong \overline{DC}$

Prove: $\overline{AC} \cong \overline{DB}$

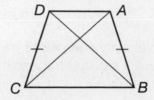

STRATEGY   Write a two-column proof.

Statements	Reasons
1. *ABCD* is an isosceles trapezoid, $\overline{AB} \cong \overline{DC}$	1. Given
2. $\angle ABC \cong \angle DCB$	2. Theorem 30.1: A trapezoid is isosceles if and only if the base angles are congruent.
3. $\overline{BC} \cong \overline{BC}$	3. Reflexive Property
4. $\triangle ABC \cong \triangle DCB$	4. SAS
5. $\overline{AC} \cong \overline{DB}$	5. CPCTC

SOLUTION   The two-column proof is shown above.

EXAMPLE 2

Prove a trapezoid is isosceles if and only if the base angles are congruent.

Prove that in all isosceles trapezoids the supplements to the base angles are also congruent.

Given: *EFGH* is an isosceles trapezoid with $\overline{EF} \cong \overline{HG}$.

Prove: $\angle F \cong \angle G$ and $\angle FEH \cong \angle H$

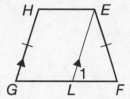

STRATEGY   Write a two-column proof.

Statements	Reasons
1. *EFGH* is an isosceles trapezoid and $\overline{EF} \cong \overline{HG}$.	1. Given
2. $\overline{EH} \parallel \overline{LG}$	2. Definition of trapezoid
3. *ELGH* is a parallelogram.	3. Definition of parallelogram
4. $\angle G \cong \angle 1$	4. Corresponding Angles Theorem
5. $\overline{HG} \cong \overline{EL}$	5. Opposite sides of a parallelogram are congruent.
6. $\overline{EF} \cong \overline{EL}$	6. Transitive Property
7. $\angle F \cong \angle 1$	7. Base angles of an isosceles triangle are congruent.
8. $\angle F \cong \angle G$	8. Transitive Property
9. $\angle F + \angle FEH = 180°$ and $\angle G + EHG = 180°$	9. Consecutive Interior Angles Theorem
10. $\angle FEH \cong \angle H$	10. Supplements of congruent angles are congruent.

SOLUTION   The two-column proof is shown above.

The **midsegment** of a trapezoid connects the midpoints of the legs. The midsegment is also called the median and the midline.

> **Theorem 30.4**: In a trapezoid, the midsegment is parallel to the bases and equal to one-half the sum of the bases.

EXAMPLE 3

Find the length of the midsegment in the diagram.

STRATEGY   Use Theorem 30.4.

STEP 1   Set up an equation.

Set the expression for the midsegment equal to one-half the sum of the bases.

$x + 10 = \frac{1}{2}(x + 1 + 3x + 5)$

STEP 2   Solve the equation.

$x + 10 = \frac{1}{2}(x + 1 + 3x + 5)$

$x + 10 = \frac{1}{2}(4x + 6)$

$x + 10 = 2x + 3$

$x + 7 = 2x$

$x = 7$

STEP 3   Substitute the value into the expression for the median to find its value.

$x + 10 = 7 + 10 = 17$

SOLUTION   **The midsegment is 17 units long.**

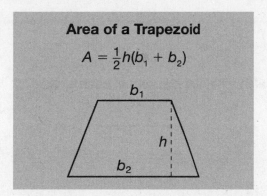

Area of a Trapezoid

$$A = \frac{1}{2}h(b_1 + b_2)$$

EXAMPLE 4

Find the area of a trapezoid with $b_1 = 12$ cm, $b_2 = 16$ cm, and $h = 9$ cm.

STRATEGY   Use the formula for area.

$$A = \frac{1}{2}h(b_1 + b_2)$$

$$A = \frac{1}{2}(9)(12 + 16)$$

$$A = \frac{1}{2}(9)(28)$$

$$A = 126$$

SOLUTION   The area of the trapezoid is 126 cm^2.

COACHED EXAMPLE

The trapezoid shown is isosceles. Find the measure of each angle.

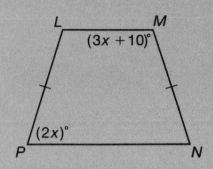

THINKING IT THROUGH

The angles marked are _____ angles.

In an isosceles trapezoid, opposite angles are _____.

_____ + _____ = 180

_____ + _____ = 180

_____ = _____

x = _____

Substitute the value you found for x to find m$\angle P$.

2 · _____ = _____

m$\angle P$ = _____

m$\angle M$ = _____

m$\angle L$ = _____

m$\angle N$ = _____

Lesson Practice

Choose the correct answer.

1. The sum of the angles formed by the same leg of a trapezoid equals

    (1) 90°.

    (2) 135°.

    (3) 180°.

    (4) 360°.

2. If one base of a trapezoid is 15 inches and the other base is 25 inches, what is the length of the median?

    (1) 20 inches

    (2) 30 inches

    (3) 40 inches

    (4) 50 inches

3. If two angles of a right trapezoid are each 90°, what is the sum of the remaining angles?

    (1) 90°

    (2) 180°

    (3) 270°

    (4) 360°

4. Suppose one diagonal of an isosceles trapezoid measured 19 feet. What is the length of the other diagonal?

    (1) 9.5 feet

    (2) 12 feet

    (3) 17 feet

    (4) 19 feet

5. One leg of an isosceles trapezoid is $3x + 5$ cm. The other leg is $5x - 1$ cm. What is the length of each leg of the isosceles trapezoid?

    (1) 3 cm

    (2) 6 cm

    (3) 9 cm

    (4) 14 cm

6. Find the area of the trapezoid with $b_1 = 23$ in. $b_2 = 29$ in., and $h = 6$ in.

    (1) 104 in.^{2}

    (2) 156 in.^{2}

    (3) 402.5 in.^{2}

    (4) 420.5 in.^{2}

7. Given isosceles trapezoid *PQRS*, what is the measure of base ∠*QPS* if the corresponding base ∠*PQR* = 130°?

   (1) 45°

   (2) 50°

   (3) 100°

   (4) 130°

8. Which of the following is not sufficient to prove a trapezoid is isosceles?

   (1) The diagonals are congruent.

   (2) The legs are congruent.

   (3) Consecutive angles are supplementary.

   (4) Base angles are congruent.

OPEN-ENDED QUESTION

9. Given: *TRAP* is a trapezoid and $\overline{TA} \cong \overline{RP}$

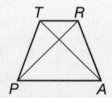

   Prove: ∠*RPA* ≅ ∠*TAP*

   _____

   _____

   _____

   _____

   _____

   _____

   _____

31 Relationships among Quadrilaterals

G.G.41

This graphic organizer shows the relationships between quadrilaterals and parallelograms.

Quadrilaterals

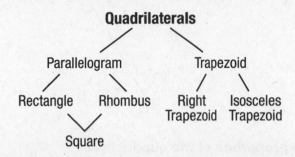

To prove that figures are special quadrilaterals you must show that they possess certain characteristics.

Quadrilateral	Properties
Trapezoid	• two sides parallel • two pairs of consecutive angles are supplementary
Parallelogram	• two pairs of parallel sides • one pair of congruent, parallel sides • diagonals bisect each other • two pairs of opposite angles are congruent • each angle is supplementary to both its consecutive angles
Rectangle	• parallelogram with four right angles • parallelogram with congruent diagonals
Rhombus	• parallelogram with four congruent sides • parallelogram with diagonals that bisect the angles • parallelogram with perpendicular diagonals
Square	• parallelogram that is both a rectangle and a rhombus

EXAMPLE 1

What type of quadrilateral is *ABCD*?

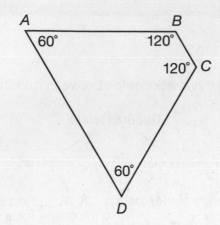

STRATEGY **Examine the properties of the quadrilateral.**

STEP 1 None of the angles are right angles so it cannot be a rectangle or a square. The angles are not equal so it cannot be a rhombus. No pairs of angles are equal so it cannot be a parallelogram.

STEP 2 The measures of consecutive angles are supplementary.

$$m\angle A + m\angle B = 180°$$

and

$$m\angle C + m\angle D = 180°$$

This means that $\overline{BC} \parallel \overline{AD}$

SOLUTION *ABCD* is a trapezoid.

Note: Upon further investigation it can be proven that the trapezoid is an isosceles trapezoid.

EXAMPLE 2

A quadrilateral has diagonals that intersect to form four congruent segments. What type of quadrilateral is it?

STRATEGY   **Draw a diagram and examine the characteristics.**

STEP 1   Draw the four congruent segments that form diagonals.

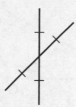

STEP 2   Connect the adjacent endpoints of the segments and label each point.

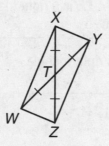

STEP 3   Classify the quadrilateral based on the diagonals.

Because the diagonals bisect each other, the quadrilateral must be a parallelogram.

$XZ = XT + TZ$ and $WY = WT + TY$

$WT = XT$ and $TY = TZ$

so

$WY = XT + TZ$ by substitution.

$WY = XZ$

The diagonals are congruent so the parallelogram is a rectangle.

SOLUTION   **The quadrilateral is a rectangle.**

COACHED EXAMPLE

In the figure, $EI = 6$, $FI = 8$ and $EF = 10$. What type of quadrilateral is *EFGH*?

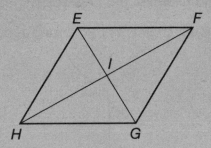

THINKING IT THROUGH

Use the Pythagorean Theorem to show that $\triangle EFI$ is a right triangle.

_____^{2} + _____^{2} = _____ + _____ = _____ = _____^{2}

Therefore, it has been proven that $\triangle EFI$ is a right triangle.

A figure with perpendicular diagonals is a _____.

EFGH is a _____.

Lesson Practice

Choose the correct answer.

1. The diagonals of a rhombus are perpendicular and bisect

    (1) 2 opposite interior sides.

    (2) 2 opposite exterior angles.

    (3) 2 adjacent angles.

    (4) 2 opposite angles.

2. Quadrilateral *QUAD* has diagonals that bisect each other, are congruent and perpendicular. What type of quadrilateral is *QUAD*?

    (1) kite

    (2) rectangle

    (3) square

    (4) rhombus

3. A rectangle has congruent diagonals. Find the length of the diagonals if one diagonal is $4x + 10$ cm and the other diagonal is $5x - 20$ cm?

    (1) 30 cm

    (2) 90 cm

    (3) 130 cm

    (4) 180 cm

4. The diagonals are always congruent in each of the following quadrilaterals except

    (1) a rectangle.

    (2) a square.

    (3) an isosceles trapezoid.

    (4) a rhombus.

5. In rectangle *ABCD*, $\overline{BD}$ and $\overline{AC}$ are diagonals and intersect at *E*. If $\overline{BD} = 84$ units and the measure of $\overline{AE} = 6y$, what is the value of *y*?

    (1) 7

    (2) 14

    (3) 28

    (4) 84

6. All of the following are always parallelograms except a

    (1) trapezoid.

    (2) rectangle.

    (3) rhombus.

    (4) square.

OPEN-ENDED QUESTION

7. In the diagram, $\triangle ABD \cong \triangle CDB$. Prove, using a two-column proof, that $ABCD$ is a parallelogram.

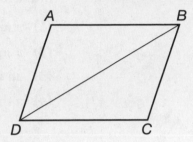

Tangents and Secants

G.G.50, G.G.53

A **tangent** is a line that touches a circle in exactly one point. The **point of tangency** is the point on the circle where the tangent line touches the circle. A **secant** is a line that intersects a circle at two points.

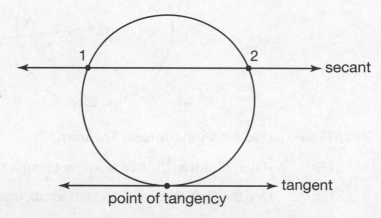

Theorem 32.1: A tangent line and a radius drawn to the point of tangency are perpendicular.

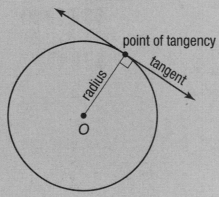

Theorem 32.2: If two segments are tangent to the same circle from the same point outside the circle, then they are congruent.

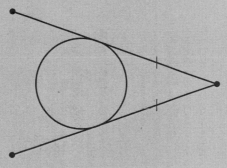

EXAMPLE 1

$\overline{NM}$ is tangent to circle L. What is the length of $\overline{LN}$?

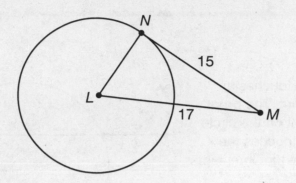

STRATEGY   **Use the Pythagorean Theorem.**

STEP 1   Recognize that $\triangle LMN$ is a right triangle because $\overline{NM}$ is perpendicular to $\overline{NL}$.

STEP 2   Use the Pythagorean Theorem to calculate the length of $\overline{LN}$.

$$a^2 + b^2 = c^2$$
$$a^2 + 15^2 = 17^2$$
$$a^2 + 225 = 289$$
$$a^2 = 64$$
$$a = 8$$

SOLUTION   **The length of $\overline{LN}$ is 8.**

EXAMPLE 2

Prove that if two segments are tangent to the same circle from the same point, then they are congruent.

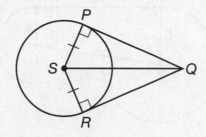

STRATEGY   **Write a two-column proof.**

Given: $\overline{QP}$ and $\overline{QR}$ are tangent to circle S.

Prove: $\overline{QP} \cong \overline{QR}$

Statements	Reasons
1. $\overline{QP}$ and $\overline{QR}$ are tangent to circle S.	1. Given
2. $\overline{PS} \cong \overline{RS}$	2. Radii in the same circle are congruent.
3. $\overline{PS} \perp \overline{QP}$ $\overline{SR} \perp \overline{QR}$	3. A tangent line and a radius drawn to the point of tangency are perpendicular.
4. $\triangle PQS$ and $\triangle RQS$ are right triangles.	4. Definition of right triangle
5. $\overline{QS} \cong \overline{QS}$	5. Reflexive property
6. $\triangle PQS \cong \triangle RQS$	6. HL
7. $\overline{QP} \cong \overline{QR}$	7. CPCTC

SOLUTION   **The two-column proof is shown above.**

EXAMPLE 3

$\overline{WX}$ is tangent to Circle Y and Circle Z. What is the length of $\overline{YZ}$?

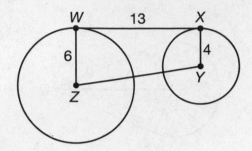

STRATEGY **Construct figures that are easy to identify.**

STEP 1 Recognize that $\overline{WX} \perp \overline{WZ}$ and $\overline{WX} \perp \overline{XY}$.

STEP 2 Draw a segment that is parallel to $\overline{WX}$. Name the segment $\overline{VY}$. $\overline{VY} \perp \overline{WZ}$. Therefore, *WVYX* is a rectangle.

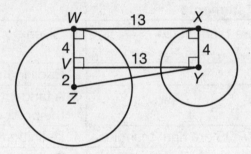

STEP 3 Use the Pythagorean Theorem to find the length of $\overline{YZ}$.

$$a^2 + b^2 = c^2$$
$$2^2 + 13^2 = c^2$$
$$4 + 169 = c^2$$
$$173 = c^2$$
$$13.15 \approx c$$

SOLUTION The length of $\overline{YZ}$ is approximately 13.15.

COACHED EXAMPLE

Circle *O* is inscribed in △*HIJ*. Find the perimeter of △*HIJ*.

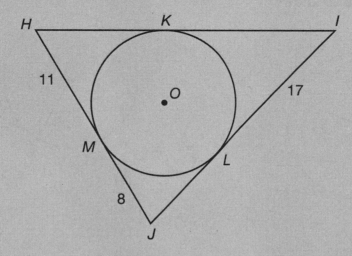

THINKING IT THROUGH

When two segments are tangent to a circle from a point outside the circle, the segments are _____.

If $\overline{MJ} \cong \overline{LJ}$ and $MJ = 8$, then $LJ =$ _____.

If $\overline{MH} \cong \overline{KH}$ and $MH = 11$, then $KH =$ _____.

If $\overline{LI} \cong \overline{KI}$ and $LI = 17$, then $KI =$ _____.

To find the perimeter of the triangle, the side lengths must be _____.

The perimeter of △*HIJ* is _____.

Lesson Practice

Choose the correct answer.

Use the following circle to answer Exercises 1 and 2.

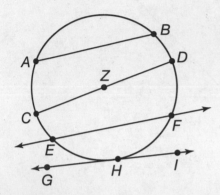

1. Which line is tangent to circle Z?

   (1) $\overline{AB}$

   (2) $\overline{CD}$

   (3) $\overleftrightarrow{EF}$

   (4) $\overleftrightarrow{GI}$

2. Which line is not a secant in circle Z?

   (1) $\overline{AB}$

   (2) $\overline{CD}$

   (3) $\overleftrightarrow{EF}$

   (4) $\overleftrightarrow{GI}$

3. $\overline{FH}$ is tangent to Circle G. Find the value of x.

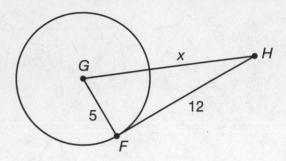

   (1) 10

   (2) 11

   (3) 12

   (4) 13

4. $\overline{RS}$ is tangent to Circle T and Circle U. What is the length of $\overline{TU}$?

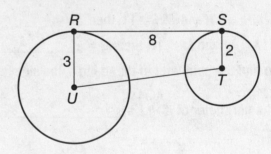

   (1) 7.94 units

   (2) 8.06 units

   (3) 8.60 units

   (4) 9.60 units

5. What is the length of $\overline{AC}$?

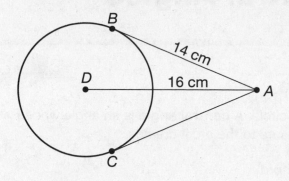

(1) 7 cm

(2) 12 cm

(3) 14 cm

(4) 16 cm

6. Circle A is inscribed in $\triangle BCD$. Find the perimeter of $\triangle BCD$.

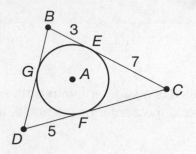

(1) 15 units

(2) 25 units

(3) 30 units

(4) 35 units

OPEN-ENDED QUESTION

7. Given: $\overline{FG}$ is tangent to Circle E at H.

$\overline{HF} \cong \overline{HG}$

Prove: $\overline{EF} \cong \overline{EG}$

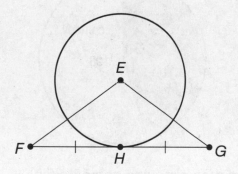

33 Chords and Central Angles

G.G.49

A **chord** is a segment with both endpoints on the circle. A **central angle** is an angle whose vertex is the center of the circle. It is equal in measure to the arc it creates.

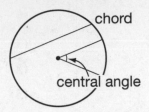

chord

central angle

EXAMPLE 1

Seven hundred fifty students were asked which type of movie they prefer to watch: action, comedy, drama, or science fiction. A circle graph was created to illustrate the results of the survey. The central angles of three of the categories are given below. How many students prefer to watch an action film?

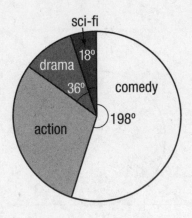

sci-fi

18°

drama

36°

comedy

198°

action

STRATEGY   **Use central angle measures and percentages.**

STEP 1   Add the given central angle measurements.

198° + 36° + 18° = 252°

STEP 2   Subtract the sum from Step 1 from 360° to find the missing angle measurement.

360° − 252° = 108°

STEP 3   Calculate the percentage of students who prefer an action film.

108 ÷ 360 = 0.3 = 30%

STEP 4   Find the number of students who prefer watching an action movie.

Multiply the percentage of students who prefer an action movie by the total number of students.

750 · 0.3 = 225

SOLUTION   **The number of students who prefer to watch an action movie is 225.**

> **Theorem 33.1**: A diameter that is perpendicular to a chord bisects the chord.

EXAMPLE 2

Prove that a diameter that is perpendicular to a chord bisects the chord.

STRATEGY   **Create a two-column proof.**

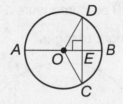

Given: $\overline{AB}$ is a diameter of circle O; $\overline{OE} \perp \overline{CD}$.

Prove: $\overline{AB}$ bisects $\overline{DC}$.

Statements	Reasons
1. $\overline{AB}$ is a diameter of circle O; $\overline{OE} \perp \overline{CD}$	1. Given
2. △OEC and △OED are right triangles.	2. Perpendicular lines form right angles and right triangles have one right angle.
3. $\overline{OD} \cong \overline{OC}$	3. Radii in the same circle are congruent.
4. $\overline{OE} \cong \overline{OE}$	4. Reflexive Property
5. △OEC ≅ △OED	5. HL
6. $\overline{DE} \cong \overline{CE}$	6. CPCTC
7. E is the midpoint of $\overline{DC}$	7. Definition of midpoint
8. $\overline{AB}$ bisects $\overline{DC}$	8. Definition of bisector

SOLUTION   **The two-column proof is shown above.**

EXAMPLE 3

Chord $\overline{FH}$ is 22 cm long and 8 cm from the center of circle E. Find the radius of circle E.

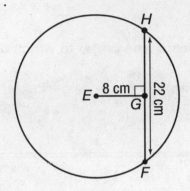

STRATEGY   **Use Theorem 33.1.**

STEP 1   Recognize that $\overline{EG}$ bisects chord $\overline{HF}$.

$\frac{1}{2} \cdot 22 = 11$, $\overline{GF}$ and $\overline{GH}$ each measure 11 cm.

STEP 2   Draw radius $\overline{EF}$ to form a right triangle.

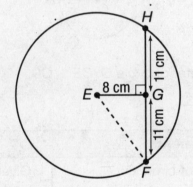

STEP 3   Use the Pythagorean Theorem to find the measure of radius $\overline{EF}$.

$$EF^2 = 8^2 + 11^2$$
$$EF^2 = 185$$
$$EF = 13.6$$

SOLUTION   **The length of a radius of circle E is 13.6 cm.**

An **arc** is a piece of a circle that connects two points on the circle. The measure of an arc is equal to the central angle that defines it. The arc from point *A* to point *B* is $\overset{\frown}{AB}$. Three points may be used to name an arc and indicate which direction about the circle the arc travels. $\overset{\frown}{ABC}$ is the arc that begins at *A* and goes through *B* to *C*.

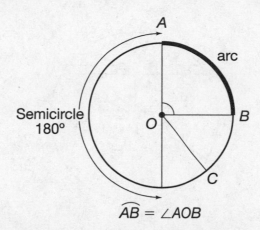

$$\overset{\frown}{AB} = \angle AOB$$

EXAMPLE 4

Find the value of *x*.

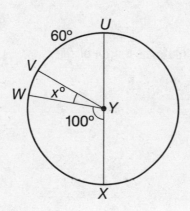

STRATEGY   Use addition and subtraction.

STEP 1   Recognize that $\overset{\frown}{UWX}$ equals 180° because it is a semicircle.

STEP 2   Recognize that m∠*UYV* = 60° because a central angle and its corresponding arc have equal measurements.

STEP 3   Add central angles *UYV* and *WYX* together, then subtract from 180° to find the measure of central angle *VYW*.

60° + 100° = 160°

180° − 160° = 20°

SOLUTION   The value of *x* is 20°.

COACHED EXAMPLE

Find the length of $\overline{OM}$.

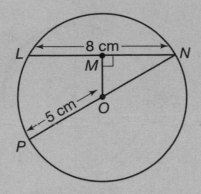

THINKING IT THROUGH

$\overline{OM}$ is the perpendicular bisector of _____.

$\overline{LM} \cong$ _____, which means both segments have a length of _____.

$\overline{PO} \cong$ _____

$\angle OMN$ is a _____ angle.

Use the Pythagorean Theorem to calculate the length of $\overline{OM}$.

The length of $\overline{OM}$ is _____.

Lesson Practice

Choose the correct answer.

1. Find the diameter of circle *R*. Round to the nearest tenth.

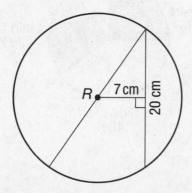

    (1) 6.3 cm

    (2) 7.1 cm

    (3) 9.4 cm

    (4) 12.2 cm

2. Find the value of *x* in circle *M*.

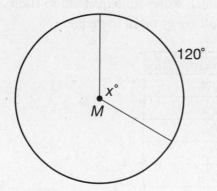

    (1) 90

    (2) 110

    (3) 120

    (4) 150

3. In a recent survey 1,500 people were asked to pick their favorite professional sport. The circle graph below illustrates the results.

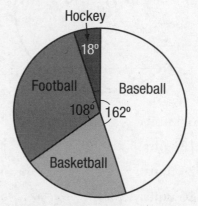

How many of the people surveyed picked basketball?

    (1) 20

    (2) 72

    (3) 259

    (4) 300

4. Find the value of *x*.

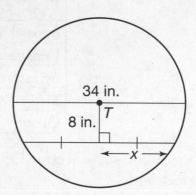

    (1) 15 in.

    (2) 16 in.

    (3) 17 in.

    (4) 18 in.

5. Find the value of *x*.

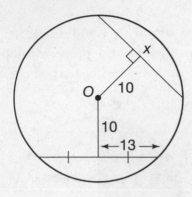

(1) 6.5 units

(2) 13 units

(3) 23 units

(4) 26 units

6. Mr. Kelton assigned each family an item to bring to a party. What is the measure of the angle that represents bringing drinks?

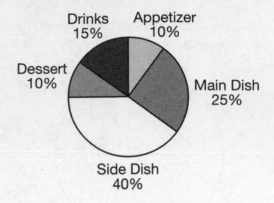

(1) 15°

(2) 36°

(3) 54°

(4) 108°

OPEN-ENDED QUESTION

7. Students at Franklin High School took a poll to find out which stores students tend to shop at when shopping for gifts for family members. The results of the poll are given in the table.

Store	Percent
Apparel Store	32%
Department Store	26%
Accessory Store	9%
Media Store	18%
Other	15%

Find the measure of each central angle of the circle. Create a circle graph to display the data. Label each angle.

34 Arc of a Circle formed by Intersecting Lines

G.G.51

An **intercepted arc** is formed by two segments intersecting with a circle.

If two lines or segments intersect in the interior of a circle, then the measure of each angle formed is half the sum of the measures of the intercepted arcs.

If two lines or segments intersect outside the circle, the measure of the angle formed is half the difference of the intercepted arcs.

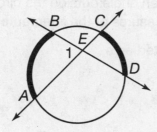

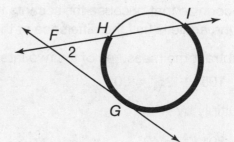

$$m\angle 1 = \tfrac{1}{2}(m\widehat{AB} + m\widehat{CD})$$

$$m\angle 2 = \tfrac{1}{2}(m\widehat{GI} - m\widehat{HG})$$

EXAMPLE 1

Find the value of *y*.

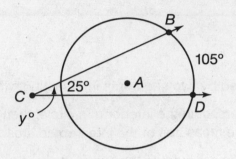

STRATEGY   **Use properties of intersecting secants.**

STEP 1   Recognize that because the secants form an angle outside the circle, the measure of the angle is half the difference of the measures of the intercepted arcs.

STEP 2   Subtract the measures of the two intercepted arcs.

105° − 25° = 80°

STEP 3   Multiply by $\frac{1}{2}$.

80° · $\frac{1}{2}$ = 40°

SOLUTION   **The value of *y* is 40°.**

EXAMPLE 2

Find the value of *y*.

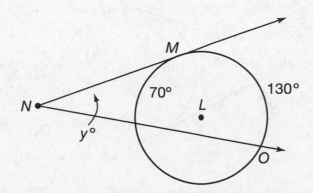

STRATEGY   Use properties of intersecting tangents and secants.

STEP 1   Recognize that because the secant and tangent form an angle outside the circle, the measure of the angle is half the difference of the measures of the intercepted arcs.

STEP 2   Subtract the measures of the two intercepted arcs.

130° − 70° = 60°

STEP 3   Multiply by $\frac{1}{2}$.

$60° \cdot \frac{1}{2} = 30°$

SOLUTION   The value of *y* is 30°.

EXAMPLE 3

Find the value of *y*.

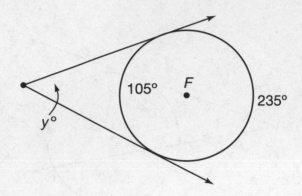

STRATEGY   **Use properties of intersecting tangents.**

STEP 1   Recognize that because two tangents form an angle outside the circle, the measure of the angle is half the difference of the measures of the intercepted arcs.

STEP 2   Subtract the measures of the two intercepted arcs.

$$235° - 105° = 130°$$

STEP 3   Multiply by $\frac{1}{2}$.

$$130° \cdot \frac{1}{2} = 65°$$

SOLUTION   **The value of *y* is 65°.**

An **inscribed** angle is an angle drawn inside a circle so that the vertex is on the circle. If an angle intersects a semicircle, then its measure is 90°.

> **Inscribed Angle Theorem**: The measure of an inscribed angle is equal to one half the measure of the intercepted arc.
>
> **Theorem 34.1:** If two angles intercept two congruent arcs, then the angles are congruent.

EXAMPLE 4

Find the values of a, b, and c.

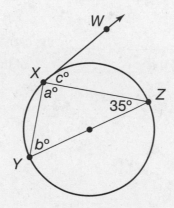

STRATEGY   Use properties of inscribed angles and intercepted arcs.

STEP 1   Find the value of a.

When an angle is inscribed in a semicircle, the angle measure is 90°. Therefore, $\angle YXZ$ measures 90°.

STEP 2   Find the value of b.

$m\angle XYZ = 180 - 90 - 35 = 55°$

STEP 3   Find the $m\widehat{XZ}$.

$2m\angle XYZ = m\widehat{XZ}$

$2 \cdot 55° = 110°$

STEP 4   Find the value of c.

$\frac{1}{2}m\widehat{XZ} = \angle WXZ$

$\left(\frac{1}{2}\right)110° = \angle WXZ = 55°$

SOLUTION $a = 90°$, $b = 55°$, $c = 55°$

COACHED EXAMPLE

Find the value of *y*.

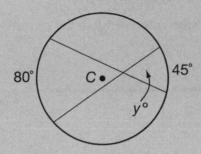

THINKING IT THROUGH

Two chords intersect _____ the circle, therefore the measure of the
angle formed by the two chords is half the _____ of the measures of the intercepted arcs.

$\frac{1}{2}($ _____ + _____ $)$

$= \frac{1}{2} \cdot$ _____

$=$ _____

The value of *y* is _____ **.**

Lesson Practice

Choose the correct answer.

1. Find the measure of $\overset{\frown}{TU}$.

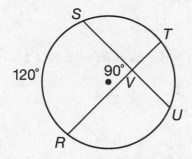

   (1) 30°

   (2) 60°

   (3) 90°

   (4) 120°

2. What is the measure of $\overset{\frown}{AB}$?

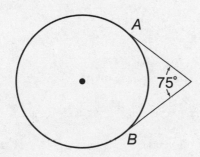

   (1) 75°

   (2) 105°

   (3) 150°

   (4) 285°

3. What is the measure of $\angle ECD$?

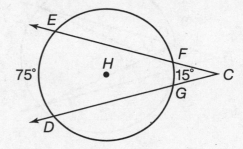

   (1) 15°

   (2) 30°

   (3) 45°

   (4) 60°

4. Find the measure of $\angle GHJ$.

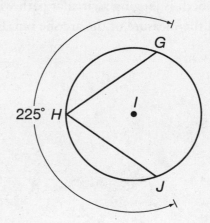

   (1) 67.5°

   (2) 112°

   (3) 135°

   (4) 225°

Use the following circle to answer Exercises 5 and 6.

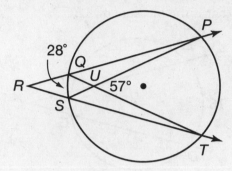

5. What is the measure of $\overset{\frown}{PT}$?

   (1) 57°

   (2) 80°

   (3) 86°

   (4) 114°

6. What is the measure of $\angle PRT$?

   (1) 14°

   (2) 29°

   (3) 57°

   (4) 114°

OPEN-ENDED QUESTION

7. Elizabeth is jogging a circular path with a radius of 200 feet. She has jogged a distance of 350 feet. Find the measure of the arc she ran. Draw a diagram.

35 Arcs of a Circle formed by Parallel Lines

When parallel lines intersect with a circle, they create two congruent arcs. Recall that parallel lines create sets of congruent and supplementary angles when they are cut by a transversal.

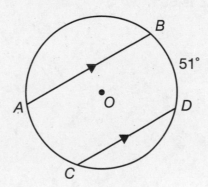

EXAMPLE 1

$\widehat{BD}$ measures 51°. What is the measure of $\widehat{AC}$?

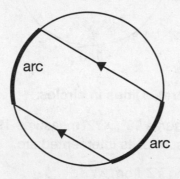

STRATEGY Use properties of parallel lines in circles.

STEP 1 Recognize that $\overline{AB}$ and $\overline{CD}$ are parallel.

STEP 2 Because parallel chords have congruent arcs, $\widehat{AC}$ is congruent to $\widehat{BD}$.

SOLUTION $\widehat{AC}$ measures 51°.

EXAMPLE 2

Find the measure of $\overset{\frown}{WZ}$.

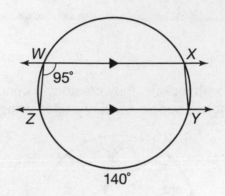

STRATEGY **Use properties of parallel lines in circles.**

STEP 1 Because $\angle XWZ$ measures 95°, $\overset{\frown}{XYZ}$ measures 190°. The measure of an inscribed angle is half the measure of its intercepted arc.

STEP 2 Find $\overset{\frown}{XY}$ by subtracting $\overset{\frown}{YZ}$ from $\overset{\frown}{XYZ}$.
$$\overset{\frown}{XY} = 190° - 140° = 50°$$

STEP 3 Because $\overset{\frown}{XY}$ is congruent to $\overset{\frown}{WZ}$, the measure of $\overset{\frown}{WZ}$ = 50°.

SOLUTION **The measure of $\overset{\frown}{WZ}$ is 50°.**

EXAMPLE 3

The following circle contains an inscribed rectangle. Find the measure of $\overset{\frown}{HIG}$.

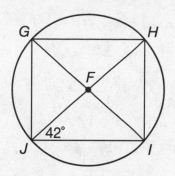

STRATEGY   **Use properties of parallel lines in circles.**

STEP 1   The measure of an inscribed angle is half the measure of its intercepted arc.

Because $\angle HJI$ measures 42°, $\overset{\frown}{HI}$ measures 84°.

STEP 2   Recall that the diagonals of a rectangle are congruent.

Because congruent chords have congruent arcs, and $\overset{\frown}{HI}$ measures 84°, then $\overset{\frown}{GJ}$ measures 84°.

STEP 3   Because $\overline{GJ}$ and $\overline{HI}$ are parallel, $\overset{\frown}{GH}$ and $\overset{\frown}{JI}$ have equal measures.

STEP 4   Recall that a circle is made up of 360°.

$\overset{\frown}{HI}$ and $\overset{\frown}{GJ}$ measure 168° together, so subtract 168° from 360°.

84° + 84° = 168°

360° − 168° = 192°

$\overset{\frown}{GH}$ and $\overset{\frown}{JI}$ measure 192° together.

STEP 5   Divide the sum of the two arcs in half to find the measure of $\overset{\frown}{GH}$ and $\overset{\frown}{JI}$.

192° ÷ 2 = 96°

STEP 6   Add the measures of $\overset{\frown}{HI}$, $\overset{\frown}{IJ}$, and $\overset{\frown}{JG}$ together to find the total measure of $\overset{\frown}{HIG}$.

84° + 84° + 96° = 264°

SOLUTION   **The measure of $\overset{\frown}{HIG}$ is 264°.**

COACHED EXAMPLE

What type of quadrilateral is the inscribed polygon?

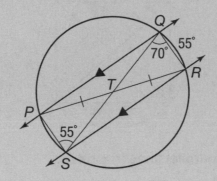

THINKING IT THROUGH

Because $\overarc{QR}$ measures 55°, then $\angle QSR$ measures _____.

$\overline{PQ}$ and $\overline{SR}$ are _____.

Alternate interior angles prove that $\angle QSR$ and $\angle$_____ are equal.

$\angle QRS$ can be calculated by subtracting the two known angles from _____ in $\triangle QRS$.

$\angle QPS$ can be calculated by subtracting the two known angles from _____ in $\triangle QPS$.

$\angle QRS =$ _____ $-$ _____ $=$ _____.

$\angle QPS =$ _____ $-$ _____ $=$ _____.

After calculating the angle measures of each corner of the quadrilateral, it can be concluded that quadrilateral *PQRS* is more specifically a _____.

Lesson Practice

Choose the correct answer.

Use the following circle to answer Exercises 1 and 2.

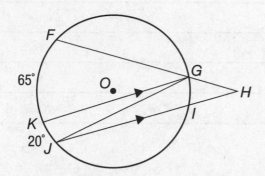

1. What is the measure of $\overarc{GI}$?

   (1) 20°

   (2) 32°

   (3) 40°

   (4) 80°

2. Which measure is the greatest?

   (1) $\overarc{GI}$

   (2) $\angle GJI$

   (3) $\angle KGJ$

   (4) $\angle FHJ$

Use the following circle to answer Exercises 3–5.

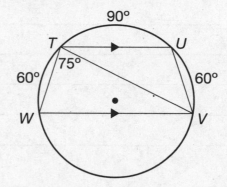

3. Find the measure of $\angle UTV$.

   (1) 20°

   (2) 30°

   (3) 60°

   (4) 75°

4. Find the measure of $\angle UVT$.

   (1) 20°

   (2) 30°

   (3) 40°

   (4) 45°

5. Name the following inscribed polygon.

   (1) parallelogram

   (2) rectangle

   (3) square

   (4) trapezoid

OPEN-ENDED QUESTION

6. If a parallelogram is inscribed in a circle, what kind of parallelogram must it be? Explain your answer.

    _____

    _____

    _____

    _____

    _____

36 Segments Intersected by a Circle

G.G.53

When two chords intersect inside a circle, the product of the segments of one chord is equal to the product of the segments of the other chord.

EXAMPLE 1

Find the length of $\overline{CF}$.

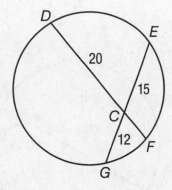

STRATEGY   **Use properties of intersecting chords in a circle.**

STEP 1   Recognize that two chords are intersecting inside the circle.

STEP 2   Recall that if two chords intersect inside a circle, the product of the segments of one chord is equal to the product of the segments of the other chord.

$$DC \cdot CF = EC \cdot CG$$
$$20 \cdot CF = 15 \cdot 12$$
$$20 \cdot CF = 180$$
$$CF = 9$$

SOLUTION   **The length of $\overline{CF}$ is 9.**

When two secants intersect outside the circle the products of their lengths and external segments are congruent. When a tangent and a secant intersect outside a circle the square of the length of the tangent = the product of the secant and external segment.

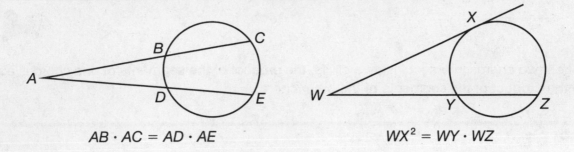

$$AB \cdot AC = AD \cdot AE \qquad\qquad WX^2 = WY \cdot WZ$$

EXAMPLE 2

Find the length of $\overline{HI}$.

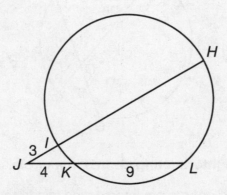

STRATEGY Use properties of secants intersecting a circle.

STEP 1 Recognize that two secants are intersecting the circle.

STEP 2 Recall that if two secants are drawn from a point outside a circle, the product of the lengths of one secant and its external segment equals the product of the lengths of the other secant and its external segment.

$$(JI + HI)\, JI = (JK + KL)\, JK$$
$$(3 + HI)3 = (4 + 9)4$$
$$(3)HI + 9 = 52$$
$$(3)HI = 43$$
$$HI = 14.\overline{3}$$

SOLUTION The length of $\overline{HI}$ is 14.$\overline{3}$.

EXAMPLE 3

Find the length of tangent $\overline{MN}$.

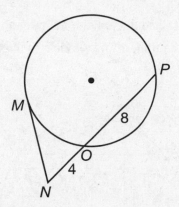

STRATEGY   Use properties of tangents and secants intersecting outside a circle.

STEP 1   Recognize that a tangent and a secant are intersecting the circle.

STEP 2   Recall that if a tangent and a secant are drawn from a point outside a circle, then the product of the lengths of the secant and its external segment equals the square of the length of the tangent.

$$(NO + OP)\, NO = MN^2$$
$$(4 + 8)4 = MN^2$$
$$48 = MN^2$$
$$MN \approx 6.9$$

SOLUTION   The length of $\overline{MN}$ is approximately 6.9 units.

COACHED EXAMPLE

The Kiln Family has a personalized stained glass window hanging above their front door to welcome everyone to their home. A diagram of the window is illustrated below. Find the radius of the circle.

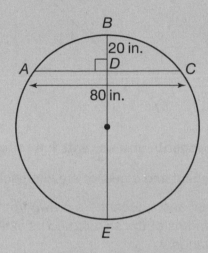

THINKING IT THROUGH

The 80-inch chord is _____ by perpendicular $\overline{BD}$, which means that $\overline{AD}$ and $\overline{DC}$ have lengths of _____.

In the diagram above, $\overline{BE}$ is not only a chord but also the _____.

Use the following formula to find the missing segment of the chord.

_____ · _____ = _____ · _____

The length of chord $\overline{DE}$ is _____.

The diameter of the circle is _____.

The radius of the circle is _____.

Lesson Practice

Choose the correct answer.

1. Find the length of $\overline{YZ}$.

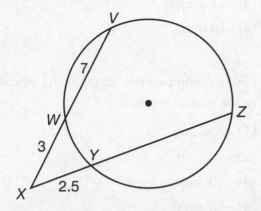

   (1) 6.5 units

   (2) 7.5 units

   (3) 9.5 units

   (4) 10.5 units

2. Find the length of $\overline{SR}$.

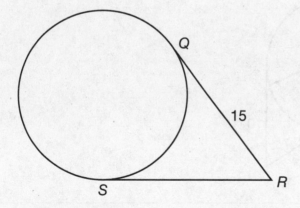

   (1) 5 units

   (2) 7.5 units

   (3) 15 units

   (4) 30 units

3. Two chords of a circle intersect. The first chord is made up of a 14-inch segment and a 9-inch segment. The second chord is made up of a 10.5-inch segment and a second segment. What is the length of the second segment of the second chord?

   (1) 13 in.

   (2) 12 in.

   (3) 11 in.

   (4) 10 in.

4. A student is to find the length of $\overline{EF}$. What equation should she use?

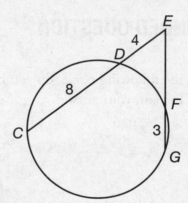

   (1) $4(8 + 4) = x(3 + x)$

   (2) $8(4 + 8) = 3(x + 3)$

   (3) $4x = 8 \cdot 3$

   (4) $4 \cdot 8 = 3 \cdot x$

Use the following circle to answer Exercises 5 and 6.

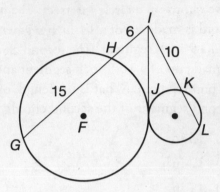

5. Find the approximate length of $\overline{IJ}$ rounded to the nearest tenth.

   (1) 8.8 units

   (2) 10.6 units

   (3) 11.2 units

   (4) 16 units

6. Find the approximate length of $\overline{KL}$ rounded to the nearest tenth.

   (1) 2.6 units

   (2) 3.4 units

   (3) 4.1 units

   (4) 4.7 units

OPEN-ENDED QUESTION

7. Using the following circle, do you need to know the value of x in order to find the value of y or vice versa? Explain your answer.

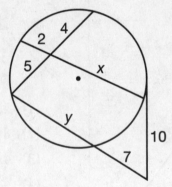

_____

_____

What are the values of x and y?

_____

Circle Proofs

 G.G.49, G.G.50, G.G.51, G.G.52, G.G.53

Recall that a proof is a convincing argument that uses deductive reasoning. Proofs can be written in various formats such as a two-column proof and a paragraph proof. The proofs used in this lesson involve circles.

EXAMPLE 1

Given: $\overline{NP}$ is tangent to circle M at O.

$\overline{ON} \cong \overline{OP}$

Prove: $\overline{MN} \cong \overline{MP}$

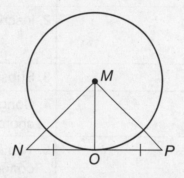

STRATEGY   Write a two-column proof.

Statements	Reasons
1. $\overline{NP}$ is tangent to circle M at O.	1. Given
2. $\overline{ON} \cong \overline{OP}$	2. Given
3. $\overline{NP} \perp \overline{MO}$	3. A tangent line and a radius drawn to the point of tangency are perpendicular.
4. $\overline{MO}$ is the perpendicular bisector of $\overline{NP}$.	4. Definition of a perpendicular bisector.
5. $\overline{MN} \cong \overline{MP}$	5. A point on the perpendicular bisector of a segment is equidistant from the endpoints of the segment.

SOLUTION   The two-column proof is shown above.

EXAMPLE 2

Given: $\overset{\frown}{KL} \cong \overset{\frown}{MN}$

Prove: $\triangle KML \cong \triangle MKN$

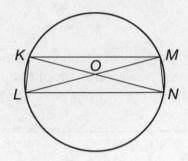

STRATEGY   Write a two-column proof.

Statements	Reasons
1. $\overset{\frown}{KL} \cong \overset{\frown}{MN}$	1. Given
2. $\angle KML = \frac{1}{2}\overset{\frown}{KL}$ and $\angle MKN = \frac{1}{2}\overset{\frown}{MN}$	2. Inscribed Angle Theorem
3. $\angle MKN \cong \angle KML$	3. Substitution Property
4. $\overline{KL} \cong \overline{MN}$	4. Congruent arcs have congruent chords.
5. $\angle KNM \cong \angle MLK$	5. Two inscribed angles that intercept congruent arcs are congruent.
6. $\triangle KML \cong \triangle MKN$	6. AAS Congruency Theorem

SOLUTION   The two-column proof is shown above.

EXAMPLE 3

Given: Chords $\overline{QS}$ and $\overline{RT}$ intersect.

Prove: $m\angle 1 = \frac{1}{2}(m\widehat{RS} + m\widehat{QT})$

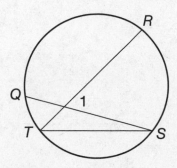

STRATEGY **Write a paragraph proof.**

STEP 1   By the Exterior Angle Theorem, $\angle 1 = m\angle RTS + m\angle QST$.

STEP 2   Because of the Inscribed Angle Theorem, $m\angle RTS = \frac{1}{2}m\widehat{RS}$ and $m\angle QST = \frac{1}{2}m\widehat{QT}$.

STEP 3   Use substitution to replace values in the first equation.

$$m\angle 1 = m\angle RTS + m\angle QST$$
$$m\angle 1 = \frac{1}{2}m\widehat{RS} + \frac{1}{2}m\widehat{QT}$$

STEP 4   Simplify using the Distributive Property.

$$m\angle 1 = \frac{1}{2}(m\widehat{RS} + m\widehat{QT})$$

SOLUTION   **The paragraph proof is shown above.**

COACHED EXAMPLE

Given: Circle *O* with diameter $\overline{PR} \perp QS$ at *T*.
Prove: $\overline{QT} \cong \overline{ST}$ and $\overarc{PQ} \cong \overarc{PS}$

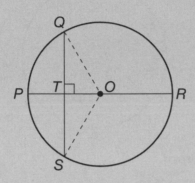

THINKING IT THROUGH

The first statement of a proof is the _____.

Which two segments are both radii? _____

Which definition makes $\angle QTO$ and $\angle STO$ right angles? _____

Which segment is shared by both triangles? _____

The two right triangles are congruent by what congruence theorem? _____

How can you prove the segments are congruent now that you know the triangles are congruent? _____

Statements	Reasons
1. $\overline{PR} \perp \overline{QS}$	1. _____
2. _____	2. Radii in the same circle are congruent.
3. $\angle QTO$ and $\angle STO$ are right angles	3. _____
4. _____	4. Reflexive Property
5. $\triangle QTO \cong$	5. _____
6. $\overline{QT} \cong \overline{ST}$	6. _____
7. $\overline{PQ} \cong \overline{PS}$	7. CPCTC
8. $\overarc{PQ} \cong \overarc{PS}$	8. _____

Lesson Practice

Choose the correct answer.

1. The first step of a proof should usually state which of the following?

   (1) segment measures

   (2) angle measures

   (3) the conclusion

   (4) the given information

2. Use the circle to complete the statement. Given that $\overline{DH}$ is a diameter and $\overline{DH} \perp \overline{EG}$, then _____ $\cong$ _____ and _____ $\cong$ _____.

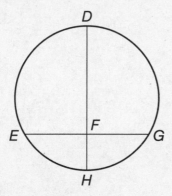

   (1) $\overline{FG}$; $\overline{FH}$; $\overarc{HG}$; $\overarc{HE}$

   (2) $\overline{FE}$; $\overline{FG}$; $\overarc{HG}$; $\overarc{HE}$

   (3) $\overline{FE}$; $\overline{FG}$; $\overarc{EH}$; $\overarc{ED}$

   (4) $\overline{FE}$; $\overline{FG}$; $\overarc{GD}$; $\overarc{GH}$

3. In a proof, one of the statements says that $\angle TUV = \frac{1}{2}\overarc{TWV}$. Which of the following reasons explains this statement?

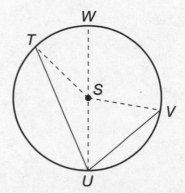

   (1) Corresponding Angles Postulate

   (2) Alternate Interior Angles Theorem

   (3) Inscribed Angle Theorem

   (4) Pythagorean Theorem

4. In the diagram, $m \parallel n$. Choose the circle theorem that would prove $\triangle LMN$ is isosceles.

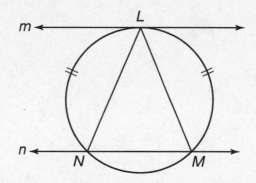

   (1) Theorem 32.1

   (2) Theorem 33.1

   (3) Theorem 34.1

   (4) Inscribed Angle Theorem

OPEN-ENDED QUESTION

5. Draw a figure that illustrates the following theorem:

   A diameter that is perpendicular to a chord bisects the chord and its arc.

Constructions

 G.G.17, G.G.18, G.G.19, G.G.20

A **compass** is a tool used for drawing arcs and circles. A **straightedge** is similar to a ruler but without the markings for measurement. It can only be used to draw a straight line. Using knowledge of circles and their properties, you can draw many geometric figures using only a compass and a straightedge. These drawings are **constructions**.

EXAMPLE 1

Construct the bisector of ∠ABC.

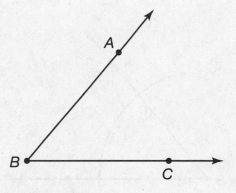

STRATEGY   **Use a compass and straightedge.**

STEP 1   Place the point of the compass on the vertex of the angle and draw an arc that intersects both sides of the angle.

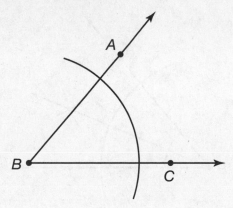

STEP 2   Place the compass point at the intersection of the arc and $\overrightarrow{BC}$. Draw a small arc in the interior of the angle.

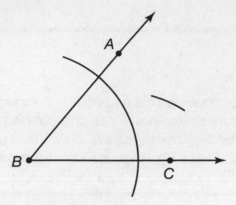

STEP 3   Keep the compass setting the same, and place the compass point at the intersection of the first arc and $\overrightarrow{BA}$. Draw an arc that intersects the arc you drew in Step 2.

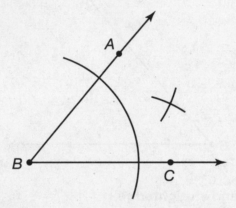

SOLUTION   Use the straightedge to draw a ray from the vertex of the angle through the intersection of the two small arcs. This ray is the angle bisector.

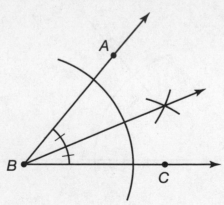

EXAMPLE 2

Construct the perpendicular bisector of $\overline{DE}$.

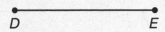

STRATEGY **Use a compass and straightedge.**

STEP 1   Place the point of the compass at D. Set the compass to be slightly larger than half the length of $\overline{DE}$. Draw an arc that extends above and below the segment.

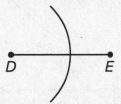

STEP 2   Without changing the compass setting, draw a similar arc from point E. Extend the arc until both arcs intersect.

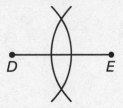

SOLUTION **Use the straightedge to draw a segment connecting the two intersections of the arcs. This segment is the perpendicular bisector.**

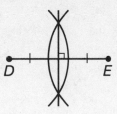

EXAMPLE 3

Construct a line parallel to $\overleftrightarrow{RS}$.

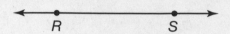

STRATEGY Construct two congruent corresponding angles.

STEP 1 Draw a line that intersects $\overleftrightarrow{RS}$ at R. Label it $\overleftrightarrow{RT}$.

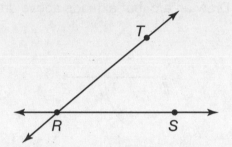

STEP 2 Draw a small arc on $\angle TRS$. Use the same compass setting to draw an arc from point T.

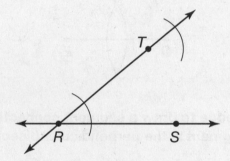

STEP 3   Place the compass point at the intersection of the first arc and $\overrightarrow{RT}$ and the pencil at the intersection on $\overrightarrow{RS}$ to measure the width of the angle at this point. Mark this width on the second arc by placing the compass point on the intersection of the second arc with $\overrightarrow{RT}$ and the pencil on the point of the second arc that is in the interior of ∠TRS.

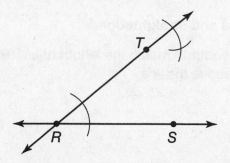

SOLUTION   Using a straightedge, draw a line that connects point *T* to the intersection of the arc you drew in Step 3 and the arc you drew in Step 2.

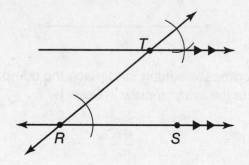

EXAMPLE 4

Construct an equilateral triangle with side length equal to $\overline{LM}$.

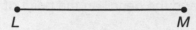

STRATEGY **Use the compass and straightedge.**

STEP 1 Set the compass width to match the length of $\overline{LM}$. Place the tip of the compass at L and draw an arc above the line.

STEP 2 Keep the same compass setting, and place the compass point at M and draw an arc that intersects the arc you drew in Step 1.

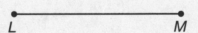

SOLUTION **Use the straightedge to draw segments from the intersection of the two arcs to L and M.**

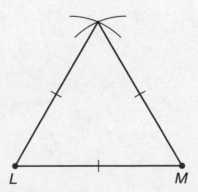

COACHED EXAMPLE

Construct a line perpendicular to line *m* through point *P*.

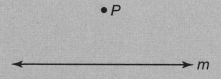

THINKING IT THROUGH

First, draw an arc centered at *P* that intersects *m* twice.

Second, draw arcs centered at each intersection so that they intersect below the line.

Finally, use a straightedge to draw a line from *P* through the intersection of the arcs below the line.

Lesson Practice

Choose the correct answer.

1. The diagram shows the construction of a line perpendicular to another line. Which is the reason that PQ is perpendicular to line n?

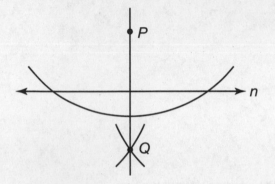

(1) P and Q are equidistant from the two points on line n.

(2) Perpendicular lines form right angles.

(3) Two congruent triangles are formed.

(4) Points equidistant from a given point form a circle.

2. Which step is not included in the procedure for constructing an equilateral triangle?

(1) Set the compass to the length of the first segment.

(2) Draw an arc centered at one endpoint above the line.

(3) Draw an arc centered at the other endpoint that intersects the first arc.

(4) Draw two intersecting arcs under the first segment.

3. The diagram shows the construction of an angle bisector. Which of the following must be true?

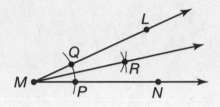

(1) $LQ = NP$

(2) $RP = RQ$

(3) $RL = RN$

(4) $MN = ML$

4. The diagram shows the construction of a line parallel to a given line. Which of the following must be true?

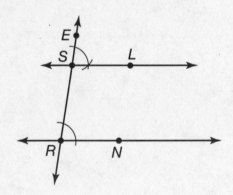

(1) $\angle ESL \cong \angle SRN$

(2) $SL = RN$

(3) $\angle ELS \cong \angle SNR$

(4) $ES = SR$

5. Which statement is always true about the intersection of two arcs centered at the endpoints of a segment?

(1) It is the center of the circle in which the arcs are contained.

(2) It is equidistant from the midpoint of the segment.

(3) It is equidistant from the endpoints of the segment.

(4) It is the midpoint of the segment.

6. Which of the following could you use to construct a 45° angle?

(1) construct parallel lines, then bisect an angle

(2) construct perpendicular lines, then bisect an angle

(3) bisect an angle, then construct perpendicular lines

(4) construct perpendicular lines, then copy a segment twice

OPEN-ENDED QUESTION

7. Copy ∠*XYZ* onto line *n*.

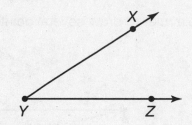

Explain the steps used in your construction.

_____

_____

_____

_____

_____

_____

39 Locus

A **locus** (plural - loci) is the set of all points that satisfy some condition. The condition usually involves a given distance from another object. A locus may be a point, line, circle, or a three–dimensional figure. The word locus means location.

EXAMPLE 1

What is the locus of points equidistant from the endpoints of $\overline{AB}$?

STRATEGY   Draw several points equidistant from the endpoints of the segment.

> STEP 1   Use a ruler to draw several points that are equidistant from the endpoints.

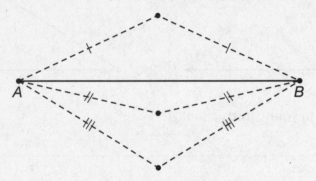

> STEP 2   Draw lines through the points you drew in Step 1.

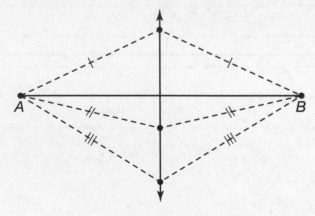

STEP 3   Examine the line and determine its relationship to $\overline{AB}$.

Use a ruler to measure the two segments created. They are congruent.

SOLUTION   **The locus of points equidistant from the endpoints of a segment is the perpendicular bisector.**

EXAMPLE 2

What is the locus of points equidistant from two intersecting lines?

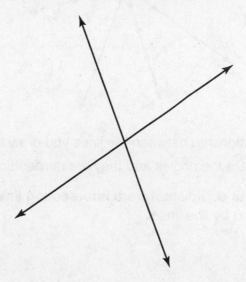

STRATEGY   **Draw several points that are equidistant from the lines.**

STEP 1   Use a ruler to draw several points that are equidistant from the sides in each angle.

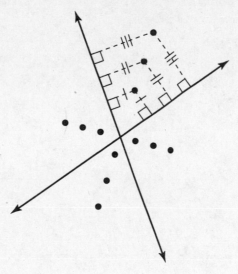

STEP 2   Draw a line through the points you found.

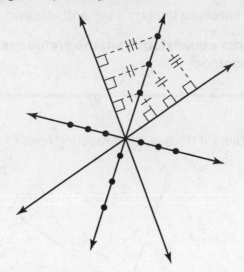

STEP 3   Examine the relationship between the lines you drew and the intersecting lines.

The lines bisect the angles and they are perpendicular.

SOLUTION   **The loci of points equidistant from intersecting lines are the angle bisectors of the angle formed by the lines.**

COACHED EXAMPLE

What is the locus of points exactly 5 units above the line $y = -2$?

THINKING IT THROUGH

Draw the line in the coordinate plane.

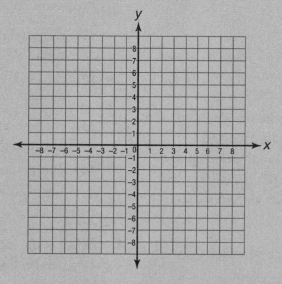

Plot several points that are exactly 5 units above the line.

Draw a line through the points you drew.

Describe the locus of points equidistant from a line. _____

_____

_____

Lesson Practice

Choose the correct answer.

1. What is the locus of points that are equidistant from the point $(-1,1)$?

    (1) a parallel line

    (2) a perpendicular line

    (3) a circle

    (4) an angle bisector

2. Wilson got lost hiking in the woods. A rescue team found his campsite and estimated that he could have traveled up to 3 miles since morning. What is the locus of points in which the rescuers should search?

    (1) a line 3 miles long

    (2) a square 3 miles long and wide

    (3) a ray 3 miles from the campsite

    (4) a circle with radius 3 miles

3. Two towns lie on a road that runs east and west and are 20 miles apart. The towns will build a road that runs north and south and is equally accessible to both towns. Which of the following should they use to construct the new road?

    (1) the perpendicular bisector of the distance between them

    (2) the circle centered at the midpoint between the towns

    (3) the bisector of the angle formed between the towns and due north

    (4) a line parallel to the road on which the towns lie

4. What is the locus of points equidistant from two parallel lines?

    (1) a parallel line

    (2) a perpendicular line

    (3) a circle

    (4) an angle bisector

5. What is the locus of points in the interior of an angle equidistant from the sides of the angle?

    (1) a parallel line

    (2) a perpendicular line

    (3) a circle

    (4) the angle bisector

6. What is the locus of points equidistant from a line?

    (1) parallel lines

    (2) perpendicular lines

    (3) a circle

    (4) an angle bisector

7. A radio broadcasting tower has a radius of 15 miles. What is the locus of the tower?

    (1) a circle

    (2) a line

    (3) a ray

    (4) a square

OPEN-ENDED QUESTION

8. A cell phone company is improving coverage between Syracuse and Utica. They wish to place 5 towers equidistant between the two cities so that they are no less than 2 miles and no more than 4 miles apart.

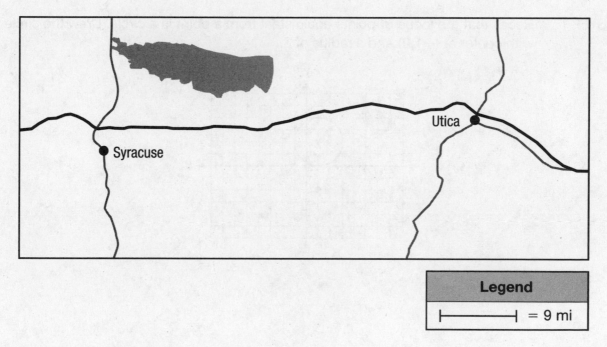

Sketch 5 points where the company can place their towers. Explain your reasoning.

40 Compound Locus

G.G 22, 23

A **compound locus** is a set of points that satisfies two or more conditions. When solving a compound locus question, find each locus separately, then find the intersection of these loci.

EXAMPLE 1

Find the number of points that are both 2 units away from the point $(-1,2)$ and equidistant between $(-1,2)$ and $(-1,-2)$.

STRATEGY   **Find each locus separately.**

STEP 1   Recall that the locus of points equidistant from a point is a circle. Draw the circle with center at $(-1,2)$ and a radius of 2.

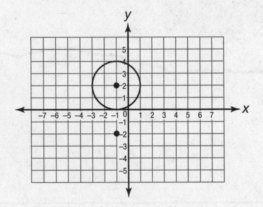

STEP 2   Graph the line connecting (−1,2) with (−1,−2). Recall that the locus of points equidistant from (−1,2) and (−1,−2) is the perpendicular bisector of the segment with those endpoints. In this case, it is the *x*-axis. Graph this on the same coordinate plane.

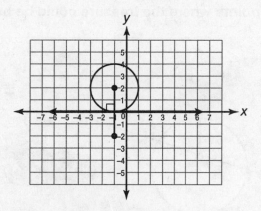

STEP 3   Find the intersection of the two loci.

SOLUTION   **The circle and the line intersect at only one point: (−1,0).**

EXAMPLE 2

A treasure map shows the location of a large rock, a tree, and a small pool. The location of a buried chest is described as being the same distance from the rock as from the pool and 3 paces from the tree. Where would you dig to find the treasure?

Legend
⊢⊣ = 1 pace

STRATEGY   **Find each locus separately.**

STEP 1   The locus of points equidistant from two points is the perpendicular bisector of the segment between them. Draw a segment connecting the rock and the pool. Then draw the segment's perpendicular bisector.

STEP 2   The locus of points equidistant from one point is a circle. Draw a circle around the tree with a radius of 3 paces.

STEP 3   Find the intersection of the two loci.

SOLUTION   **There are two points where the treasure could be buried.**

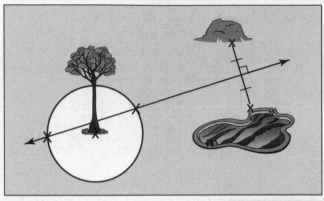

Legend
⊢ = 1 pace

COACHED EXAMPLE

Points *A* and *B* are 5 units apart. How many points are there that are equidistant from both points and 7 units from *B*?

THINKING IT THROUGH

The locus of points equidistant from two points is a _____.

Draw the two points 5 units apart and draw the locus.

The locus of points 7 units from *B* is a _____.

Draw the locus on the same drawing as above.

The intersection of the loci is _____.

How many points are there that are equidistant from both points and 7 units from *B*?_____

Lesson Practice

Choose the correct answer.

1. A city wants to place a water tower exactly 10 miles from the center of the city. They want it to be built alongside the New York Thruway, which runs 8 miles from the center of the city. How many locations can the city planner choose from?

   (1) 0
   (2) 1
   (3) 2
   (4) 3

2. Two parallel lines are 5 units apart. D is a point on one line. How many locus points are there that are equidistant from both lines and 2 units from D?

   (1) 0
   (2) 1
   (3) 2
   (4) 3

3. Point B is 10 units from line m. How many points are there that are 4 units from B and 7 units from the line?

   (1) 0
   (2) 1
   (3) 2
   (4) 3

4. What is the number of points located 5 units from the origin and 2 units from the line $x = 10$?

   (1) 0
   (2) 1
   (3) 2
   (4) 3

5. Which point is included in the locus of points located 3 units from the x-axis and 7 units from the y-axis?

   (1) $(0,7)$
   (2) $(0,3)$
   (3) $(7,3)$
   (4) $(3,7)$

6. The distance between parallel lines m and n is 15 units. Point L is on line n. How many points are equidistant from the parallel lines and 12 units from L?

   (1) 0
   (2) 1
   (3) 2
   (4) 3

7. Which of the following shows the locus of points 4 units from (2,3) and 3 units from (4,−3)?

(1)

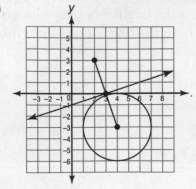

(3)

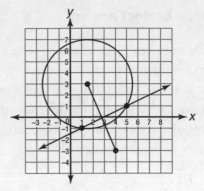

(2)

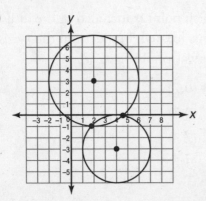

(4)

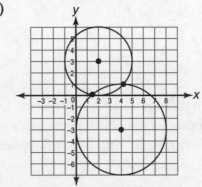

OPEN-ENDED QUESTION

8. The diagram at the right shows the location of Tyrese's school, work, and his favorite park. He wants to live an equal distance between work and school and no more than 5 miles away from the park. Where should Tyrese look for an apartment?

Show your work.

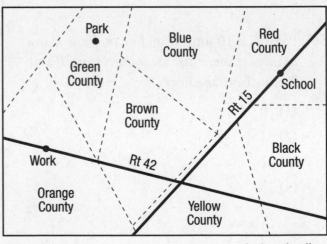

Lesson

41 Slope of a Line

 G.G.62, G.G.63, G.G.64, G.G.65

Every line has a slant, or **slope**, that is the same everywhere on the line.

The slope of a line is a ratio: $\frac{\text{change in } y}{\text{change in } x}$, or $\frac{\text{rise}}{\text{run}}$.

For the line containing the points P_1 (x_1, y_1) and P_2 (x_2, y_2):

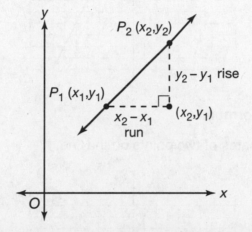

Slope Formula

$$\text{slope} = \frac{y_2 - y_1}{x_2 - x_1}$$

- The slope of a horizontal line is 0.

- The slope of a vertical line is undefined.

EXAMPLE 1

What is the slope of this line?

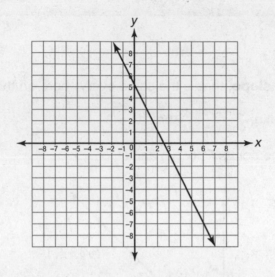

STRATEGY   Use the slope formula.

STEP 1   Find the coordinates of two points on the line.

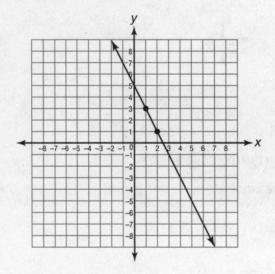

$$x_1 = 2, y_1 = 1, \text{ and } x_2 = 1, y_2 = 3$$

STEP 2   Substitute values into the slope formula.

$$m = \frac{y_2 - y_1}{x_2 - x_1}$$

$$m = \frac{3 - 1}{1 - 2}$$

$$m = \frac{2}{-1}$$

$$m = -2$$

SOLUTION   The slope of the line is −2.

Parallel lines have the same slope. Perpendicular lines have slopes that are negative reciprocals (product of the reciprocal and −1).

EXAMPLE 2

Are the segments below parallel, perpendicular, or neither?

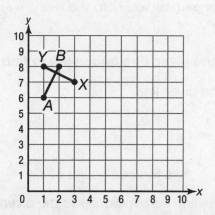

STRATEGY: **Find the slope of each segment. Then analyze the slopes.**

STEP 1: Find the slope of $\overline{AB}$.

$$m = \frac{y_2 - y_1}{x_2 - x_1}$$

$x_1 = 1, y_1 = 6,$ and $x_2 = 2, y_2 = 8$

$$m = \frac{8 - 6}{2 - 1}$$

$$m = \frac{2}{1} = 2$$

STEP 2: Find the slope of $\overline{XY}$.

$$m = \frac{y_2 - y_1}{x_2 - x_1}$$

$x_1 = 3, y_1 = 7,$ and $x_2 = 1, y_2 = 8$

$$m = \frac{8 - 7}{1 - 3}$$

$$m = \frac{1}{-2}$$

STEP 3: Identify the relationship between the two slopes.

If the slopes are the same, the segments are parallel.

If the slopes are negative reciprocals, the segments are perpendicular.

If the slopes have no relationship, the segments are neither.

The slopes are 2 and $\frac{1}{-2}$. They are negative reciprocals.

SOLUTION: **The segments are perpendicular.**

To find the equation of a line, use a point and a slope. Find the equation of a line parallel or perpendicular to any given line through a given point by substituting values for *m*, *x*, and *y* into the slope-intercept formula of a line, $y = mx + b$.

EXAMPLE 3

Find the equation of the line that is perpendicular to the line $y = \frac{-1}{3}x + 1$ and passes through the point $(-1,2)$.

STRATEGY   **Find the slope of the given line and use it to find the new equation.**

STEP 1   Find the slope of the given line.

$$y = mx + b$$
$$y = \frac{-1}{3}x + b$$
$$m = \frac{-1}{3} \qquad m \text{ is the coefficient of } x.$$

STEP 2   Find the slope of the line perpendicular to the given line.

Perpendicular lines have opposite reciprocal slopes.

$$-\frac{1}{\frac{-1}{3}} = -(-3) = 3$$

STEP 3   Use the slope you found in Step 2 and the given point to find the *y*-intercept of the new equation.

$2 = 3(-1) + b$ Substitute $(-1,2)$ and 3 into $y = mx + b$ for x, y, and m.
$2 = -3 + b$ Multiply.
$5 = b$ Add 3 to both sides.

STEP 4   Substitute the values you found in Steps 2–3 into the slope-intercept equation.

$m = 3$
$b = 5$
$y = mx + b$ Substitute the values into the equation.

SOLUTION   The equation is $y = 3x + 5$.

COACHED EXAMPLE

Find the equation of the line parallel to $y = \frac{1}{2}x - 5$ that passes through the point $(-4, 1)$.

THINKING IT THROUGH

The slopes of parallel lines are _____.

The slope of the given line is _____.

The slope of the parallel line is _____.

Substitute the slope and coordinates into $y = mx + b$ to solve for the y-intercept.

_____ = _____ · _____ $+ b$

_____ = _____ $+ b$

_____ $= b$

The equation of the line is $y =$ _____ $x +$ _____.

Lesson Practice

Choose the correct answer.

1. Which statement is true about the slope of the line that passes through the points $(5,2)$ and $(-1,2)$?

   (1) The slope is undefined.

   (2) The slope is 0.

   (3) The slope is 3.

   (4) The slope is $\frac{1}{3}$.

2. What is the slope of the line that passes through the points $(2,3)$ and $(-1,4)$?

   (1) -3

   (2) $\frac{-1}{3}$

   (3) $\frac{1}{3}$

   (4) 3

3. What is the slope of this line?

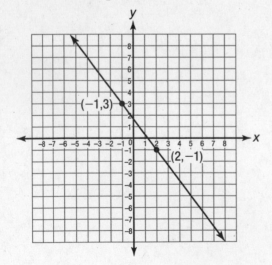

   (1) $\frac{-4}{3}$

   (3) $\frac{3}{4}$

   (2) $\frac{-3}{4}$

   (4) $\frac{4}{3}$

4. Find the equation of the line parallel to $y = -2x + 3$ that passes through the point $(-1,1)$.

   (1) $y = -2x - 1$

   (2) $y = 2x + 1$

   (3) $y = \frac{-1}{2}x + 1$

   (4) $y = \frac{1}{2}x - 1$

5. Which of the following is perpendicular to the line $y = \frac{3}{4}x - 1$?

   (1) $y = \frac{-3}{4}x + 1$

   (2) $y = \frac{3}{4}x - 2$

   (3) $y = \frac{-4}{3}x + 5$

   (4) $y = \frac{4}{3}x - 7$

6. Which statement is true about the slope of lines?

   (1) In $y = mx + b$, the slope is b.

   (2) The rise is the change in the y values.

   (3) It equals $\frac{\text{run}}{\text{rise}}$.

   (4) Perpendicular lines have the same slope.

7. A line segment contains the points $(4,8)$ and $(2,-4)$. What is the slope of a line segment perpendicular to the given line segment?

   (1) $\frac{-3}{2}$

   (2) $\frac{-1}{6}$

   (3) $\frac{2}{3}$

   (4) 6

OPEN-ENDED QUESTION

8. A quadrilateral has vertices $A(1,4)$, $B(5,0)$, $C(2,-3)$, and $D(-2,1)$. Find the slopes of each side of the quadrilateral. Determine if the quadrilateral is a parallelogram, rectangle, or both.

   A. Determine if the quadrilateral is a parallelogram. Explain your reasoning.

   _____

   _____

   _____

   B. Determine if the quadrilateral is a rectangle. Explain your reasoning.

   _____

   _____

   _____

Distance and Midpoint Formulas

G.G.66, G.G.67

The diagram shows how the Pythagorean Theorem is used to justify the **distance formula**.

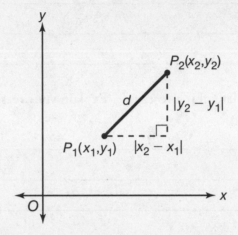

To find the distance d between any two points P_1 (x_1, y_1) and P_2 (x_2, y_2), use the following formula:

Distance Formula
If d is the distance between two points (x_1, y_1) and (x_2, y_2) in the coordinate plane, then $d = \sqrt{(x_2 - x_1)^2 + (y_2 - y_1)^2}$.

EXAMPLE 1

Find the distance between $M(-3,2)$ and $N(5,4)$.

STRATEGY   **Use the Distance Formula.**

STEP 1   Identify (x_1,y_1) and (x_2,y_2).

The first point, M, has coordinates $(-3,2)$.

$x_1 = -3$ and $y_1 = 2$

The second point, N, has coordinates $(5,4)$.

$x_2 = 5$ and $y_2 = 4$

STEP 2   Substitute the values into the formula and simplify.

$d = \sqrt{(x_2 - x_1)^2 + (y_2 - y_1)^2}$

$d = \sqrt{(5 - (-3))^2 + (4 - 2)^2}$ Subtract within the parentheses.

$d = \sqrt{(8)^2 + (2)^2}$ Simplify exponents.

$d = \sqrt{64 + 4}$ Add.

$d = \sqrt{68}$

$d = \sqrt{4 \cdot 17}$ $\sqrt{4 \cdot 17} = \sqrt{4} \cdot \sqrt{17}$

$d = 2\sqrt{17}$

SOLUTION   The distance between M and N is $2\sqrt{17}$.

The length of a segment is the same as the distance between its endpoints. Use the Distance Formula to find the length of any segment.

EXAMPLE 2

Find the length of $\overline{AB}$.

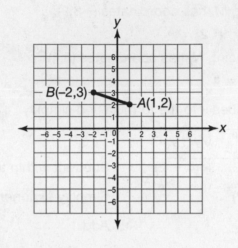

STRATEGY **Use the Distance Formula.**

STEP 1 Identify (x_1,y_1) and (x_2,y_2).

The first point, A, has coordinates (1,2).

$x_1 = 1$ and $y_1 = 2$

The second point, B, has coordinates (−2,3).

$x_2 = -2$ and $y_2 = 3$

STEP 2 Substitute the values into the formula and simplify.

$d = \sqrt{(x_2 - x_1)^2 + (y_2 - y_1)^2}$

$d = \sqrt{(-2 - 1))^2 + (3 - 2)^2}$

$d = \sqrt{(-3)^2 + (1)^2}$ Subtract within the parentheses.

$d = \sqrt{9 + 1}$ Simplify exponents.

$d = \sqrt{10}$ Add.

SOLUTION **The length of $\overline{AB}$ is $\sqrt{10}$ units.**

The **midpoint** of a segment is the point that separates the segment into two parts of equal length. The midpoint of a segment can be found using the following formula:

Midpoint of a segment

$$(x_m, y_m) = \left(\frac{x_1 + x_2}{2}, \frac{y_1 + y_2}{2}\right)$$

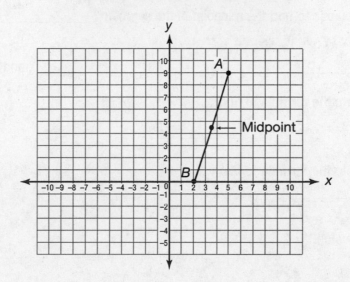

EXAMPLE 3

The endpoints of a side of a triangle are (4,5) and (−2,1). At what coordinate will a median of the triangle intersect this side?

STRATEGY   Find the midpoint of the segment using the Midpoint Formula.

   STEP 1   Identify x_1, x_2, y_1, and y_2.

   The first point has coordinates (4,5).

   So, $x_1 = 4$ and $y_1 = 5$.

   The second point has coordinates (−2,1).

   So, $x_2 = -2$ and $y_2 = 1$.

   STEP 2   Substitute the values into the formula and simplify.

$$\left(\frac{x_1 + x_2}{2}, \frac{y_1 + y_2}{2}\right)$$

$$= \left(\frac{4 + (-2)}{2}, \frac{5 + 1}{2}\right)$$

$$= \left(\frac{2}{2}, \frac{6}{2}\right)$$

$$= (1,3)$$

SOLUTION   The midpoint of the segment, or the intersection of the median, is (1,3).

COACHED EXAMPLE

A segment has an endpoint at (3,6) and a midpoint at (−2,4). What are the coordinates of the other endpoint of the segment?

THINKING IT THROUGH

What formula should you use to find the midpoint of the segment? _____

What are the values of x_1, x_2, y_1, y_2, x_m, and y_m?

$x_1 = $ _____, $x_2 = $ _____, $y_1 = $ _____, $y_2 = $ _____, $x_m = $ _____, and $y_m = $ _____

Substitute the values into the formula to find x_2.

$$\frac{\underline{\hspace{2cm}} + \underline{\hspace{2cm}}}{2} = \underline{\hspace{2cm}}$$

Substitute the values into the formula to find y_2.

$$\frac{\underline{\hspace{2cm}} + \underline{\hspace{2cm}}}{2} = \underline{\hspace{2cm}}$$

Solve for the unknown value.

So, the other endpoint of the segment is _____.

Lesson Practice

Choose the correct answer.

Use this figure for Exercises 1 and 2.

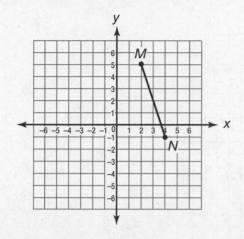

1. What is the length of $\overline{MN}$?

   (1) $\sqrt{34}$

   (2) $6\sqrt{2}$

   (3) $2\sqrt{10}$

   (4) 8

2. What is the midpoint of $\overline{MN}$?

   (1) $(0.5, 3.5)$

   (2) $(1, 3)$

   (3) $(3, 2)$

   (4) $(3.5, 1.5)$

3. A segment has an endpoint at $(-3, -1)$ and a midpoint at $(3, 2)$. What is the coordinate of the other endpoint of the segment?

   (1) $(0, 0.5)$

   (2) $(-2, -1.5)$

   (3) $(6, 5)$

   (4) $(9, 5)$

4. A segment has an endpoint at $(-3, -7)$ and a midpoint at $(-7, -2)$. What is the distance from the midpoint to the other end of the segment?

   (1) $\sqrt{28}$

   (2) $\sqrt{41}$

   (3) $\sqrt{97}$

   (4) $\sqrt{181}$

5. What is the midpoint of a line segment with endpoints $(-3, -8)$ and $(-4, -5)$?

   (1) $(-6, -4)$

   (2) $(-5.5, -4.5)$

   (3) $(-3.5, -6.5)$

   (4) $(-3, -5)$

6. Which line segments are congruent?

   I $\overline{AB}$ with $A(3, 2)$ and $B(6, 6)$

   II $\overline{RS}$ with $R(-2, 7)$ and $S(3, 7)$

   III $\overline{UV}$ with $U(-1, -1)$ and $V(-1, 4)$

   (1) I and II only

   (2) I and III only

   (3) II and III only

   (4) I, II, and III

7. The endpoints of a diameter of a circle are at $(-4, -5)$ and $(6, -3)$. What are the coordinates of the center of the circle?

   (1) $(2, -8)$

   (2) $(2, -4)$

   (3) $(1, -4)$

   (4) $(1, -2)$

OPEN-ENDED QUESTION

8. While creating a blueprint, Ushad drew a triangle on a grid with the points $A(2,1)$, $B(5,4)$ and $C(0,3)$ as its vertices.

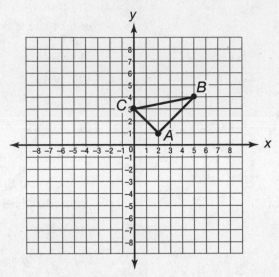

A. Use the Distance Formula to find the length of each side of the triangle. Show all of your work.

_____

_____

_____

B. Use the Slope Formula to show that $\overline{CA}$ and $\overline{BA}$ are perpendicular. Explain your answer.

_____

_____

_____

C. Classify the triangle by the length of its sides and its angles. Explain your answer.

_____

_____

_____

_____

Perpendicular Bisector of a Line Segment

G.G.68

The **perpendicular bisector** of a line segment is perpendicular to the segment at its midpoint. Any point on the perpendicular bisector of a segment is equidistant from the endpoints of the segment.

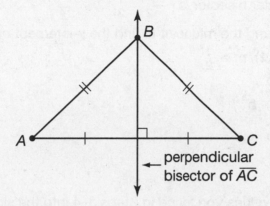

perpendicular
bisector of $\overline{AC}$

EXAMPLE 1

Find the equation of the perpendicular bisector of the segment with endpoints $A(1,-3)$ and $B(-3,5)$.

STRATEGY   **Find the midpoint of the segment, and then find the equation through that point.**

STEP 1   Use the midpoint formula to find the midpoint of the segment.

$(x_1, y_1) = (1, -3)$

$(x_2, y_2) = (-3, 5)$

$x_m = \dfrac{x_1 + x_2}{2}$ $y_m = \dfrac{y_1 + y_2}{2}$

$x_m = \dfrac{1 + (-3)}{2}$ $y_m = \dfrac{-3 + 5}{2}$

$x_m = \dfrac{-2}{2}$ $y_m = \dfrac{2}{2}$

$x_m = -1$ $y_m = 1$

The midpoint is $(-1, 1)$.

STEP 2   Find the slope of the segment.

$$m = \frac{y_2 - y_1}{x_2 - x_1}$$

$$m = \frac{5 - (-3)}{-3 - 1}$$

$$m = \frac{8}{-4}$$

$$m = -2$$

STEP 3   Find the slope of the perpendicular line.

Because perpendicular lines have opposite reciprocal slopes, the slope of the perpendicular bisector is $-\frac{1}{-2} = \frac{1}{2}$.

STEP 4   Use the slope and the midpoint to find the *y*-intercept of the line.

$$(x,y) = (-1,1), m = \frac{1}{2}$$

$$y = mx + b$$

$$1 = \frac{1}{2}(-1) + b$$

$$1 = \frac{-1}{2} + b$$

$$\frac{3}{2} = b$$

STEP 5   Substitute the values you found in Steps 3–4 into the slope-intercept formula.

SOLUTION   The equation of the perpendicular bisector is $y = \frac{1}{2}x + \frac{3}{2}$.

In order for a point to be on the perpendicular bisector of a segment, it must meet all the properties of the perpendicular bisector. However, instead of proving all the properties are true, showing that the point is equidistant from the endpoints is sufficient to prove it is on the perpendicular bisector.

EXAMPLE 2

Show that the point $(-3,-2)$ is on the perpendicular bisector of the segment with endpoints $(4,-3)$ and $(2,3)$.

STRATEGY   **Show that the point is equidistant from the endpoints.**

STEP 1   Find the distance from $(-3,-2)$ to $(4,-3)$.

$$(x_1,y_1) = (-3,-2)$$

$$(x_2,y_2) = (4,-3)$$

$$\sqrt{(x_2 - x_1)^2 + (y_2 - y_1)^2}$$

$$\sqrt{(4 - (-3)^2 + (-3 - (-2))^2}$$

$$\sqrt{7^2 + (-1)^2}$$

$$\sqrt{49 + 1}$$

$$\sqrt{50}$$

STEP 2   Find the distance from $(-3,-2)$ to $(2,3)$.

$$(x_1,y_1) = (-3,-2)$$

$$(x_2,y_2) = (2,3)$$

$$\sqrt{(x_2 - x_1)^2 + (y_2 - y_1)^2}$$

$$\sqrt{(2 - (-3))^2 + (3 - (-2))^2}$$

$$\sqrt{5^2 + 5^2}$$

$$\sqrt{25 + 25}$$

$$\sqrt{50}$$

SOLUTION   **Because the point $(-3,-2)$ is equidistant from the endpoints of the segment, it is on the perpendicular bisector of the segment.**

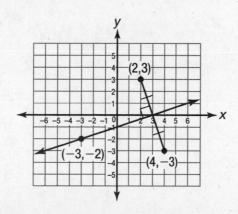

COACHED EXAMPLE

Find the equation of the perpendicular bisector of the segment with endpoints (−5,−1) and (1,3).

THINKING IT THROUGH

Find the midpoint of the segment.

$$x_m = \frac{\rule{1cm}{0.4pt} + \rule{1cm}{0.4pt}}{2}$$

$$y_m = \frac{\rule{1cm}{0.4pt} + \rule{1cm}{0.4pt}}{2}$$

$$(x_m, y_m) = (\rule{2cm}{0.4pt}, \rule{2cm}{0.4pt})$$

Find the slope of the segment.

$$\text{slope} = \frac{\rule{1cm}{0.4pt} - \rule{1cm}{0.4pt}}{\rule{1cm}{0.4pt} - \rule{1cm}{0.4pt}}$$

$$= \frac{\rule{1cm}{0.4pt}}{\rule{1cm}{0.4pt}}$$

$$= \frac{\rule{1cm}{0.4pt}}{\rule{1cm}{0.4pt}}$$

Find the slope perpendicular to the segment.

$$-\left(\frac{1}{\rule{1cm}{0.4pt}}\right) = \rule{2cm}{0.4pt}$$

Substitute the values you found into the slope-intercept equation and solve for *b*.

$$\rule{2cm}{0.4pt} = \rule{2cm}{0.4pt} \cdot \rule{2cm}{0.4pt} + b$$

$$\rule{2cm}{0.4pt} = \rule{2cm}{0.4pt} + b$$

$$\rule{2cm}{0.4pt} = b$$

Replace the slope and *y*-intercept to write the equation.

$$y = \rule{2cm}{0.4pt}x + \rule{2cm}{0.4pt}$$

Lesson Practice

Choose the correct answer.

1. Find the equation of the perpendicular bisector of the segment with endpoints $L(-3,1)$ and $M(3,3)$.

   (1) $y = \frac{1}{3}x - \frac{2}{3}$

   (2) $y = -3x + 6$

   (3) $y = \frac{1}{3}x + 2$

   (4) $y = -3x + 2$

2. What is the slope of the perpendicular bisector of a segment with slope $\frac{-5}{4}$?

   (1) $\frac{-5}{4}$

   (2) $\frac{-4}{5}$

   (3) $\frac{4}{5}$

   (4) $\frac{5}{4}$

3. Which of the following points is on the perpendicular bisector of the segment with endpoints $(-2,-1)$ and $(6,-5)$?

   (1) $(6,4)$

   (2) $(1,-5)$

   (3) $(1,-4)$

   (4) $(4,0)$

4. Which method would not work to show a point is on the perpendicular bisector of a segment?

   (1) find the equation of the perpendicular bisector and see if the point satisfies the equation

   (2) show that the distances between the point and the endpoints of the segment are equal

   (3) find the midpoint between the point and the segment

   (4) show that the slope of the line through the point and the midpoint of the segment and the slope of the segment are negative reciprocals

5. Which of the following formulas is *not* used to find the equation of the perpendicular bisector of a segment?

   (1) Distance Formula

   (2) Midpoint Formula

   (3) Slope Formula

   (4) Slope-Intercept Formula

OPEN-ENDED QUESTION

6.   Find the equation of the perpendicular bisector of the segment shown below. Show all your work. Graph the equation and show that it is a perpendicular bisector of the segment.

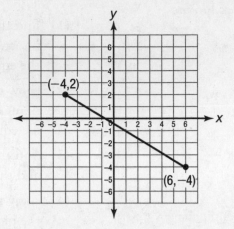

44 Properties of Triangles and Quadrilaterals

G.G.69

You can use the Slope, Distance and Midpoint Formulas to prove that a given figure is a specific polygon or to identify properties of the figure. For example, you can show that a triangle is isosceles with the Distance Formula. The formulas can also be used to investigate properties of a polygon. In **coordinate geometry**, properties of figures can be shown in a more concrete way than in Euclidean geometry.

EXAMPLE 1

Find the coordinates and length of the midsegment parallel to $\overline{AB}$.

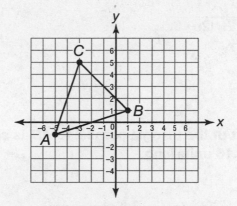

STRATEGY   **Find the endpoints of the midsegment.**

STEP 1   Find the midpoint of $\overline{BC}$.

$(x_1,y_1) = (1,1)$

$(x_2,y_2) = (-3,5)$

$$x_m = \frac{x_1 + x_2}{2} \qquad\qquad y_m = \frac{y_1 + y_2}{2}$$

$$x_m = \frac{1 + (-3)}{2} \qquad\qquad y_m = \frac{1 + 5}{2}$$

$$x_m = \frac{-2}{2} \qquad\qquad y_m = \frac{6}{2}$$

$$x_m = -1 \qquad\qquad y_m = 3$$

The midpoint of $\overline{BC}$ is $(-1,3)$.

STEP 2   Find the midpoint of $\overline{AC}$.

$(x_1, y_1) = (-5, -1)$

$(x_2, y_2) = (-3, 5)$

$x_m = \dfrac{x_1 + x_2}{2}$ $y_m = \dfrac{y_1 + y_2}{2}$

$x_m = \dfrac{-5 + (-3)}{2}$ $y_m = \dfrac{-1 + 5}{2}$

$x_m = \dfrac{-8}{2}$ $y_m = \dfrac{4}{2}$

$x_m = -4$ $y_m = 2$

The midpoint of $\overline{AC}$ is $(-4, 2)$.

STEP 2   Use the Distance Formula to find the length of the midsegment.

$(x_1, y_1) = (-1, 3)$

$(x_2, y_2) = (-4, 2)$

$\sqrt{(x_2 - x_1)^2 + (y_2 - y_1)^2}$

$\sqrt{(-4 - (-1))^2 + (2 - 3)^2}$

$\sqrt{(-3)^2 + (-1)^2}$

$\sqrt{9 + 1}$

$\sqrt{10}$

SOLUTION   **The midsegment of the triangle parallel to $\overline{AB}$ is shown in the graph. It is approximately 3.16 units long.**

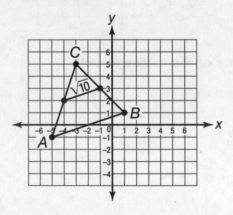

EXAMPLE 2

Find the relationship between the diagonals of the rectangle.

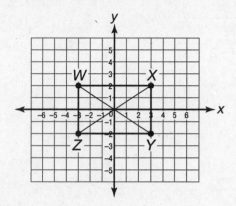

STRATEGY   Use the Distance and Slope Formulas.

STEP 1   Find the length of $\overline{WY}$.

$$(x_1, y_1) = (-3, 2)$$

$$(x_2, y_2) = (3, -2)$$

$$\sqrt{(x_2 - x_1)^2 + (y_2 - y_1)^2}$$

$$\sqrt{(3 - (-3))^2 + (-2 - 2)^2}$$

$$\sqrt{(6)^2 + (-4)^2}$$

$$\sqrt{36 + 16}$$

$$\sqrt{52}$$

$$2\sqrt{13}$$

STEP 2   Find the slope of $\overline{WY}$.

$$m = \frac{y_2 - y_1}{x_2 - x_1}$$

$$= \frac{-2 - 2}{3 - (-3)}$$

$$= \frac{-4}{6}$$

$$= \frac{-2}{3}$$

STEP 3   Find the length of $\overline{XZ}$.

$(x_1, y_1) = (3,2)$

$(x_2, y_2) = (-3, -2)$

$\sqrt{(x_2 - x_1)^2 + (y_2 - y_1)^2}$

$\sqrt{(-3 - 3)^2 + (-2 - 2)^2}$

$\sqrt{(-6)^2 + (-4)^2}$

$\sqrt{36 + 16}$

$\sqrt{52}$

$2\sqrt{13}$

STEP 4   Find the slope of $\overline{XZ}$.

$m = \dfrac{y_2 - y_1}{x_2 - x_1}$

$m = \dfrac{-2 - 2}{-3 - 3}$

$m = \dfrac{-4}{-6}$

$m = \dfrac{2}{3}$

SOLUTION   **The diagonals are congruent. They have opposite signs. They are neither parallel nor perpendicular.**

COACHED EXAMPLE

Identify the type of quadrilateral shown in the graph.

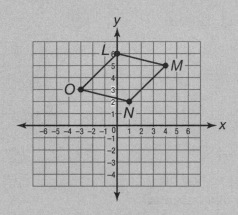

THINKING IT THROUGH

Find the slope and length of $\overline{LM}$

$(x_1,y_1) = ($ _____ , _____ $)$

$(x_2,y_2) = ($ _____ , _____ $)$

slope = _____ ; length = _____

Find the slope and length of $\overline{MN}$

$(x_1,y_1) = ($ _____ , _____ $)$

$(x_2,y_2) = ($ _____ , _____ $)$

slope = _____ ; length = _____

Find the slope and length of $\overline{NO}$

$(x_1,y_1) = ($ _____ , _____ $)$

$(x_2,y_2) = ($ _____ , _____ $)$

slope = _____ ; length = _____

Find the slope and length of $\overline{LO}$

$(x_1,y_1) = ($ _____ , _____ $)$

$(x_2,y_2) = ($ _____ , _____ $)$

slope = _____ ; length = _____

The quadrilateral is a _____ .

Lesson Practice

Choose the best answer.

1. Which formula is not used to identify properties of triangles and quadrilaterals?

   (1) Distance Formula

   (2) Slope Formula

   (3) Midpoint Formula

   (4) Slope-intercept Formula

2. Which formula would you use to find the endpoints of the midsegment of a trapezoid?

   (1) Distance Formula

   (2) Slope Formula

   (3) Midpoint Formula

   (4) Slope-intercept Formula

3. Which formula would you use to determine if a parallelogram is a rectangle?

   (1) Pythagorean Theorem

   (2) Slope Formula

   (3) Midpoint Formula

   (4) Slope-intercept Formula

4. What formula could you use to show that all three sides of an equilateral triangle are congruent?

   (1) Distance Formula

   (2) Slope Formula

   (3) Midpoint Formula

   (4) Slope-intercept Formula

5. What is the relationship between the diagonals of the square?

   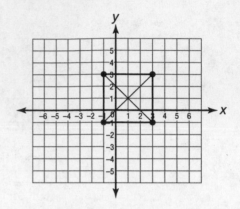

   (1) congruent

   (2) parallel

   (3) perpendicular

   (4) congruent and perpendicular

6. What type of triangle is shown in the graph?

   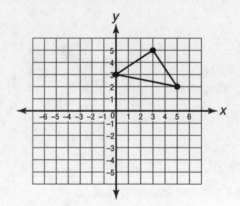

   (1) isosceles

   (2) equilateral

   (3) obtuse

   (4) scalene

OPEN-ENDED QUESTION

7. On the coordinate plane parallelogram *RSTU* has vertices $R(-3,1)$, $S(3,4)$, and $T(6,2)$. Find the coordinates of *U* and explain how you know that it forms a parallelogram.

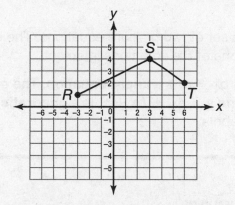

45 Solving Systems of Equations Graphically

A **system of equations** is a set of two or more equations. The **solution** to the system is the set of points where the graphs of the equations intersect.

An equation in the form $y = bx + c$, is a **linear equation**. The graph of a linear equation is a line. An equation in the form $y = ax^2 + bx + c$, is a **quadratic equation**. The graph of a quadratic equation is a parabola.

EXAMPLE 1

Solve this system of equations graphically.

$y = 2x - 2$
$y = x^2 - x - 6$

STRATEGY   Complete tables of values, and graph each equation.

STEP 1   Complete the tables of values.

x	y = 2x − 2	y
−2	$y = 2(-2) - 2$	−6
−1	$y = 2(-1) - 2$	−4
0	$y = 2(0) - 2$	−2
1	$y = 2(1) - 2$	0
2	$y = 2(2) - 2$	2

x	y = x² − x − 6	y
−2	$y = (-2)^2 - (-2) - 6$	0
−1	$y = (-1)^2 - (-1) - 6$	−4
0	$y = (0)^2 - 0 - 6$	−6
0.5	$y = (0.5)^2 - 0.5 - 6$	−6.25
1	$y = (1)^2 - 1 - 6$	−6
2	$y = (2)^2 - 2 - 6$	−4
3	$y = (3)^2 - 3 - 6$	0

STEP 2   Graph the ordered pairs on the same coordinate plane.

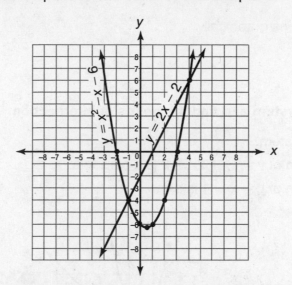

STEP 3   Find the points where the two graphs intersect.

The graphs appear to intersect at $(-1, -4)$ and $(4, 6)$.

STEP 4   Substitute the solutions into the equations to check your answer.

$$-4 \overset{?}{=} 2(-1) - 2 \qquad\qquad -4 \overset{?}{=} (-1)^2 - (-1) - 6$$

$$-4 \overset{?}{=} -2 - 2 \qquad\qquad -4 \overset{?}{=} 1 + 1 - 6$$

$$-4 = -4 \checkmark \qquad\qquad -4 \overset{?}{=} 2 - 6$$

$$-4 = -4 \checkmark$$

$$6 \overset{?}{=} 2(4) - 2 \qquad\qquad 6 \overset{?}{=} (4)^2 - (4) - 6$$

$$6 \overset{?}{=} 8 - 2 \qquad\qquad 6 \overset{?}{=} 16 - 4 - 6$$

$$6 = 6 \checkmark \qquad\qquad 6 \overset{?}{=} 12 - 6$$

$$6 = 6 \checkmark$$

SOLUTION   The solutions to the system of equations are $(-1, -4)$ and $(4, 6)$.

EXAMPLE 2

Solve this system of equations graphically.

$y = x^2 + 2x + 2$

$y = x - 1$

STRATEGY **Graph the system and find the points of intersection.**

STEP 1   Graph the system.

The graph of $y = x^2 + 2x + 2$ is a parabola.

The graph of $y = x - 1$ is a line.

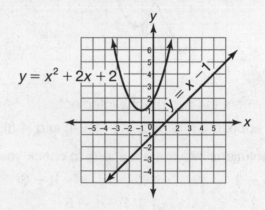

STEP 2   Identify the points of intersection.

These graphs do not intersect.

SOLUTION **There is no solution to this system of equations.**

COACHED EXAMPLE

Solve the system of equations graphically.

$y = 3$
$y = -x^2 + 3$

THINKING IT THROUGH

Graph the equations on the same coordinate plane.

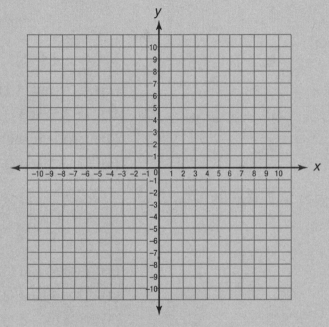

Identify the point(s) where the graphs intersect. _____

The solution to the system is _____.

Lesson Practice

Choose the correct answer.

Solve each system.

1. $y = 2x - 3$
   $y = x^2 - 6x + 9$

   (1) $(1,5)$ and $(7,17)$
   (2) $(-2,1)$ and $(-6,9)$
   (3) $(2,1)$ and $(6,9)$
   (4) There is no solution.

2. $y = 3x + 3$
   $y = x^2 + x - 12$

   (1) $(5,11)$ and $(-3,9)$
   (2) $(-3,-6)$ and $(5,18)$
   (3) $(3,12)$ and $(-5,-12)$
   (4) There is no solution.

3. $y = x + 4$
   $y = x^2 + 4x + 4$

   (1) $(-3,1)$ and $(0,4)$
   (2) $(-2,0)$ and $(0,4)$
   (3) $(-2,0)$
   (4) There is no solution.

4. $y = 2x - 1$
   $y = x^2 + 4x$

   (1) $(1,3)$
   (2) $(-1,-3)$
   (3) $(-3,-1)$
   (4) There is no solution.

5. $y = x^2 - 6$
   $y = 2x - 3$

   (1) $(-3,3)$ and $(3,3)$
   (2) $(0,-5)$ and $(3,3)$
   (3) $(-1,-5)$ and $(0,-6)$
   (4) $(-1,-5)$ and $(3,3)$

6. $y = -2x - 1$
   $y = x^2 - 4$

   (1) $(3,7)$ and $(1,1)$
   (2) $(-3,5)$ and $(1,-3)$
   (3) $(-3,-5)$ and $(-1,-3)$
   (4) $(3,-7)$ and $(-1,1)$

7. $y = x + 4$
   $y = x^2 + 2x - 8$

   (1) $(-4,8)$ and $(3,1)$
   (2) $(1,-5)$ and $(8,16)$
   (3) $(-4,0)$ and $(3,7)$
   (4) $(4,0)$ and $(-3,7)$

8. $y = x - 4$
   $y = x^2 - 1$

   (1) $(-2,2)$
   (2) $(3,7)$
   (3) $(0,-1)$
   (4) There is no solution.

OPEN-ENDED QUESTION

9. Graph the system of equations on the grid below.

$y = x^2 - 7x + 12$

$y = -x + 4$

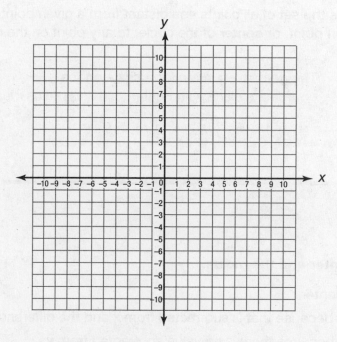

What is the solution of the system? Explain your answer.

_____

_____

_____

_____

_____

46 Equations of Circles

A **circle** is defined as the set of all points equidistant from a given point on a plane. The distance from a given point, or **center** of the circle, to any point on the circle is the **radius**.

> The general equation for a circle on the coordinate plane with center (h,k) and radius r is:
>
> $$(x - h)^2 + (y - k)^2 = r^2$$

EXAMPLE 1

Graph $(x - 6)^2 + y^2 = 9$.

STRATEGY Find the center and the radius.

> STEP 1   Find the center.
>
> > h is 6, because that is subtracted from x and the difference is squared.
> >
> > k is 0, because there is no value subtracted from y.
> >
> > The center is (6,0).
>
> STEP 2   Find the radius.
>
> > $r^2 = 9$, so $r = 3$. (You do not need to include -3 as a solution here. Why?)
>
> STEP 3   Graph the circle.
>
> > Plot (6,0). Then plot points 3 units above and below, and to the left and right of the center.

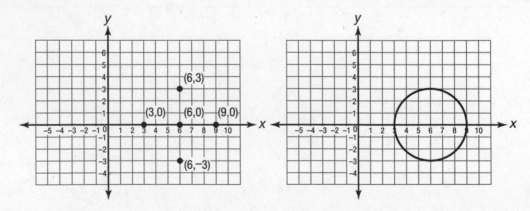

SOLUTION   Draw the circle.

Sometimes you may be given an equation in a different form. In these cases, you will have to complete the square to get the equation in (h,k) form.

EXAMPLE 2

Find the center and the radius of the circle formed by this equation:

$x^2 + y^2 - 4x + 8y + 5 = 0$

STRATEGY **Complete the square twice to put the equation in (h,k) form.**

STEP 1 Put the constant on the right side of the equation.

$x^2 + y^2 - 4x + 8y = -5$

STEP 2 Complete the square for x and y.

$(x^2 - \textbf{4}x + \underline{\hspace{1cm}}) + (y^2 + \textbf{8}y + \underline{\hspace{1cm}}) = -5 + \underline{\hspace{1cm}} + \underline{\hspace{1cm}}$

For each blank, the number will be the square of half the coefficient of the first-degree term.

$\frac{1}{2}(-4) = -2, -2^2 = 4 \qquad \frac{1}{2}(8) = 4, 4^2 = 16$

$(x^2 - 4x + 4) + (y^2 + 8y + 16) = -5 + 4 + 16$

Factor each quadratic equation and simplify.

$(x - 2)^2 + (y + 4)^2 = 15$

STEP 3 Find the center and radius.

SOLUTION **The center is (2,−4). The radius is $\sqrt{15}$.**

COACHED EXAMPLE

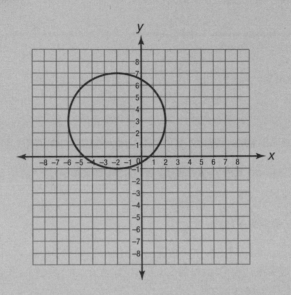

Find the equation for the circle graphed above. Then find the *x*-intercepts.

THINKING IT THROUGH

What is the center of the circle? _____

What is the radius of the circle? (Hint: count from the center to the edge vertically or horizontally.)

_____

Write an equation for the circle in (*h,k*) form. _____

To find an *x*-intercept, what must equal 0? _____

Substitute 0 in the equation and solve. To solve use the quadratic formula $x = \dfrac{-b \pm \sqrt{b^2 - 4ac}}{2a}$

_____

_____

_____

_____

Name the two intercepts. _____ **and** _____

Lesson Practice

Choose the correct answer.

1. Which equation represents the circle shown?

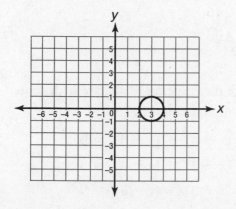

    (1) $x^2 + (y - 3)^2 = 1$

    (2) $(x - 3)^2 + y^2 = 1$

    (3) $x^2 + y^2 = 9$

    (4) $(x + 3)^2 + y^2 = 9$

2. What are the coordinates of the center of the circle $x^2 + y^2 + 8x - 2y + 6 = 0$?

    (1) $(-4, 1)$

    (2) $(-4, 2)$

    (3) $(4, 2)$

    (4) $(4, 3)$

3. Which of the following could be the graph of $x^2 + y^2 + 2x - 6y + 1 = 0$?

    (1)

    (2)

    (3)

    (4)

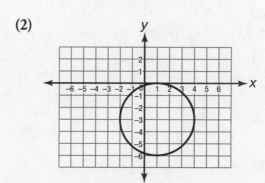

4. Which equation represents the circle shown?

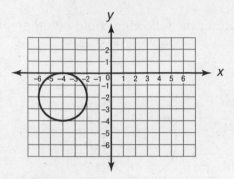

(1) $(x - 4)^2 + (y - 2)^2 = 2$

(2) $(x + 4)^2 + (y + 2)^2 = 2$

(3) $(x - 4)^2 + (y - 2)^2 = 4$

(4) $(x + 4)^2 + (y + 2)^2 = 4$

5. What is the radius of the circle

$$x^2 + y^2 + 10x - 4y + 5 = 0?$$

(1) $\sqrt{5}$

(2) $\sqrt{11}$

(3) $2\sqrt{6}$

(4) $4\sqrt{6}$

6. Write the equation of a circle with center $(-3,4)$ and radius 3.

(1) $(x - 4)^2 + (y + 3)^2 = 9$

(2) $(x - 3)^2 + (y + 4)^2 = 3$

(3) $(x + 3)^2 + (y - 4)^2 = 9$

(4) $(x + 3)^2 + (y - 4)^2 = 3$

7. Which of the following could be the graph of $x^2 + y^2 - 4x + 2y - 4 = 0$?

(1)

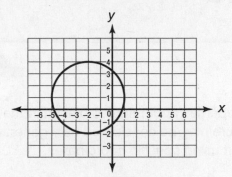

(3)

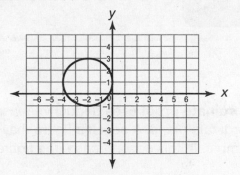

(2)

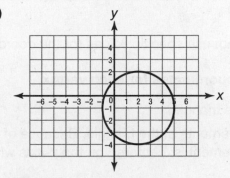

(4)

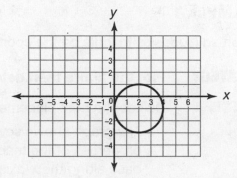

OPEN-ENDED QUESTION

8. A circle has a diameter with endpoints $(-2,3)$ and $(2,1)$. Find the center and radius of the circle and use them to write the equation of the circle.

_____

_____

_____

Coordinate Proofs

Coordinate proof uses deductive reasoning to prove properties about figures placed in the coordinate plane. A specific figure may be given, or a figure may be considered as a general example. In this second case, the figure is plotted without using specific coordinates.

EXAMPLE 1

Graph an isosceles triangle on the coordinate plane. Do not use specific values for the coordinates.

STRATEGY **Plot the figure then determine variable coordinates for each vertex.**

STEP 1 Decide the best placement of the isosceles triangle.

The graph shows several possible placements of the triangle. Because of the properties of the triangle, the best placement is: centered on the *y*-axis with the base along the *x*-axis.

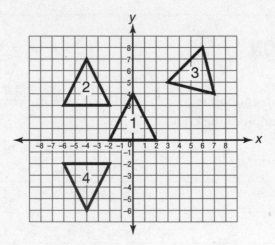

STEP 2 Find coordinates for each vertex.

Because the base is on the *x*-axis, the *y*-coordinate of both base vertices will be 0. Let the *x*-coordinate of the point on the right be *a*. Because the triangle is centered on the *y*-axis, the *x*-coordinate of the point on the left is −*a*. Let the height of the triangle be *b*. The top vertex is on the *y*-axis making its *x*-coordinate 0.

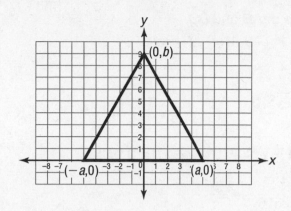

SOLUTION   The graph shows an isosceles triangle that can represent any isosceles triangle.

The Distance Formula, Slope Formula and Midpoint Formula are used in a coordinate proof to prove properties about the length and shape of figures. Different combinations of these formulas are used to prove figures have specific characteristics.

Formulas for Coordinate Proof	
Distance Formula	$d = \sqrt{(x_2 - x_1)^2 + (y_2 - y_1)^2}$
Slope Formula	$m = \dfrac{y_2 - y_1}{x_2 - x_1}$
Midpoint Formula	$(x_m, y_m) = \left(\dfrac{x_1 + x_2}{2}, \dfrac{y_1 + y_2}{2}\right)$

EXAMPLE 2

Prove that the figure below is a parallelogram.

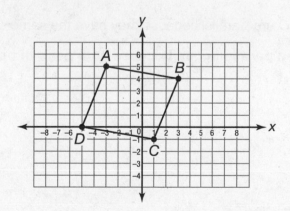

STRATEGY   Use formulas to show that both pairs of opposite sides are parallel.

STEP 1   Find the slope of $\overline{AB}$ and $\overline{DC}$.

$\overline{AB}$: $(x_1, y_1) = (-3, 5)$ $(x_2, y_2) = (3, 4)$

$m = \dfrac{y_1 - y_2}{x_1 - x_2}$

$m = \dfrac{5 - 4}{-3 - 3}$

$m = -\dfrac{1}{6}$

$\overline{DC}$: $(x_1, y_1) = (-5, 0)$ $(x_2, y_2) = (1, -1)$

$m = \dfrac{y_2 - y_1}{x_2 - x_1}$

$m = \dfrac{-1 - 0}{1 - (-5)}$

$m = \dfrac{-1}{6}$

$\overline{AB}$ and $\overline{DC}$ are parallel because they have the same slope.

STEP 2   Find the slope of $\overline{AD}$ and $\overline{BC}$.

$\overline{AD}$: $(x_1, y_1) = (-3, 5)$ $(x_2, y_2) = (-5, 0)$

$m = \dfrac{y_1 - y_2}{x_1 - x_2}$

$m = \dfrac{5 - 0}{-3 - (-5)}$

$m = \dfrac{5}{2}$

$\overline{BC}$: $(x_1, y_1) = (3, 4)$ $(x_2, y_2) = (1, -1)$

$m = \dfrac{y_2 - y_1}{x_2 - x_1}$

$m = \dfrac{-1 - 4}{1 - 3}$

$m = \dfrac{-5}{-2}$

$m = \dfrac{5}{2}$

$\overline{AD}$ and $\overline{BC}$ are parallel because they have the same slope.

SOLUTION   **The figure is a parallelogram because both pairs of opposite sides are parallel.**

EXAMPLE 3

Show that the quadrilateral formed by connecting the midpoints of each side of a rectangle is a rhombus.

STRATEGY **Show that the sides of the figure are congruent.**

STEP 1   Graph the rectangle and the quadrilateral formed by connecting the midpoints of the sides. Find the coordinates of each vertex.

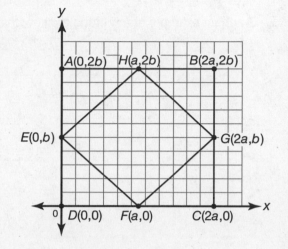

STEP 2   Find the length of each segment.

$\overline{EH}$: $(x_1, y_1) = (0, b)$ $(x_2, y_2) = (a, 2b)$

$d = \sqrt{(x_2 - x_1)^2 + (y_2 - y_1)^2}$

$d = \sqrt{(a - 0)^2 + (2b - b)^2}$

$d = \sqrt{(-a)^2 + b^2}$

$d = \sqrt{a^2 + b^2}$

$\overline{GH}$: $(x_1, y_1) = (2a, b)$ $(x_2, y_2) = (a, 2b)$

$d = \sqrt{(x_2 - x_1)^2 + (y_2 - y_1)^2}$

$d = \sqrt{(a - 2a)^2 + (2b - b)^2}$

$d = \sqrt{(-a)^2 + b^2}$

$d = \sqrt{a^2 + b^2}$

$\overline{GF}$: $(x_1, y_1) = (2a, b)$ $(x_2, y_2) = (a, 0)$

$d = \sqrt{(x_2 - x_1)^2 + (y_2 - y_1)^2}$

$d = (a - 2a)^2 + (0 - b)^2$

$d = \sqrt{(-a)^2 + (-b)^2}$

$d = \sqrt{a^2 + b^2}$

$\overline{EF}$: $(x_1, y_1) = (0, b)$ $(x_2, y_2) = (a, 0)$

$d = \sqrt{(x_2 - x_1)^2 + (y_2 - y_1)^2}$

$d = \sqrt{(a - 0)^2 + (0 - b)^2}$

$d = \sqrt{(-a)^2 + (-b)^2}$

$d = \sqrt{a^2 + b^2}$

SOLUTION **The figure has all four sides congruent; therefore, it is a rhombus.**

COACHED EXAMPLE

Show that the diagonals of a rectangle bisect each other.

THINKING IT THROUGH

Graph an example of a rectangle; show that the midpoint of the diagonals is the same.

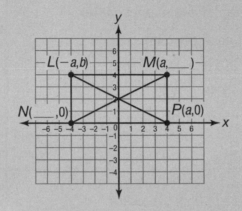

Find the midpoint of the diagonal $\overline{MN}$.

$\overline{MN}: (x_1, y_1) = ($ _____ , _____ $)$

$(x_2, y_2) = ($ _____ , _____ $)$

$(x_m, y_m) = \left(\dfrac{x_1 + x_2}{2}, \dfrac{y_1 + y_2}{2} \right)$

$= \left(\dfrac{\underline{\qquad} + \underline{\qquad}}{2}, \dfrac{\underline{\qquad} + \underline{\qquad}}{2} \right)$

$= ($ _____ , _____ $)$

Find the midpoint of the diagonal $\overline{LP}$.

$\overline{LP}: (x_1, y_1) = ($ _____ , _____ $)$

$(x_2, y_2) = ($ _____ , _____ $)$

$(x_m, y_m) = \left(\dfrac{x_1 + x_2}{2}, \dfrac{y_1 + y_2}{2} \right)$

$= \left(\dfrac{\underline{\qquad} + \underline{\qquad}}{2}, \dfrac{\underline{\qquad} + \underline{\qquad}}{2} \right)$

$= ($ _____ , _____ $)$

The diagonals of a rectangle meet at the midpoint of both lines, therefore _____
_____ .

Lesson Practice

Choose the correct answer.

1. Which of the following formulas could not be used to prove that a figure is a parallelogram?

   (1) Slope Formula

   (2) Midpoint Formula

   (3) Distance Formula

   (4) Slope-Intercept Formula

2. Which of the following formulas would be used to prove that the diagonals of a square are perpendicular?

   (1) Distance Formula

   (2) Slope Formula

   (3) Midpoint Formula

   (4) Slope-Intercept Formula

3. An isosceles trapezoid is placed in the coordinate plane. What are the coordinates of Y?

   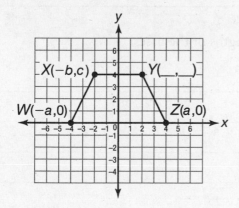

   (1) $(-b, -c)$

   (2) (a, b)

   (3) (b, c)

   (4) (a, c)

4. Use coordinate geometry to classify the figure below.

   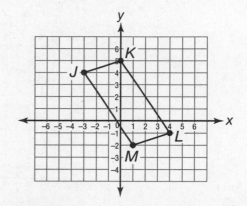

   (1) rectangle

   (2) square

   (3) rhombus

   (4) parallelogram

5. A parallelogram is placed in the coordinate plane as shown. What are the coordinates of S?

   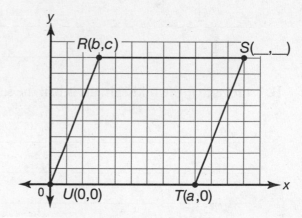

   (1) $(a + b, c)$

   (2) $(b + c, a)$

   (3) $(a + b, 0)$

   (4) $(b + c, 0)$

OPEN-ENDED QUESTION

6. Place a trapezoid in the coordinate plane. Label the coordinates of the endpoints of the trapezoid and of its midsegment.

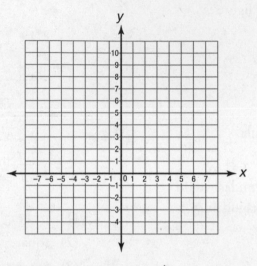

A. Prove that the midsegment of a trapezoid is parallel to the bases.

_____

_____

_____

_____

_____

B. Prove that the midsegment is half the sum of the lengths of the bases.

_____

_____

_____

_____

_____

Reflections and Symmetry

 G.G.54, G.G.55, G.G.56, G.G.61

Transformations are operations that move or re-size a figure in the coordinate plane. The original figure is the **preimage**. A transformed figure is the **image** of the preimage.

There are three types of transformations that preserve size. These are: reflections, rotations and translations. They are all called isometries because an **isometry** is a transformation that preserves size. After point A goes through a transformation, the image of the point is named A', read "A prime".

A **reflection** uses a line in the plane, the **line of reflection**, like a mirror to create an image of a figure. The preimage and image are equidistant from the line of reflection. The notation for a reflection is r_k where k is the line of reflection.

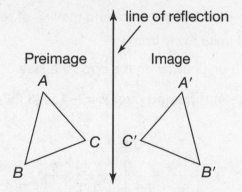

EXAMPLE 1

Reflect the point $(3, -4)$ over the line $y = -1$.

STRATEGY Find the point that is the same distance from $y = -1$ as $(3, -4)$.

STEP 1 Graph the point and the line of reflection.

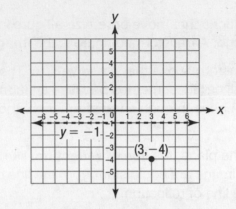

STEP 2 Find the distance between the point and the line of reflection.

The point is 3 units away from $y = -1$.

STEP 3 Locate the point 3 units away on the opposite side of the line of reflection.

SOLUTION The image of $(3, -4)$ reflected over $y = -1$ is $(3, 2)$.

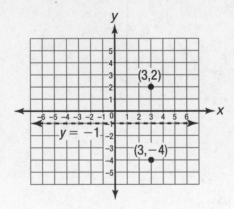

A **mapping rule** defines how the coordinates of the figure will change after the transformation. Mapping rules for the most common reflections are shown in the table below.

Reflection	Mapping Rule
$r_{y=x}$	$(x,y) \rightarrow (y,x)$
$r_{x\text{-axis}}$	$(x,y) \rightarrow (x,-y)$
$r_{y\text{-axis}}$	$(x,y) \rightarrow (-x,y)$

EXAMPLE 2

Draw $\overline{AB}$ with $A(-1,4)$ and $B(1,-2)$ after the transformation $r_{y=x}$.

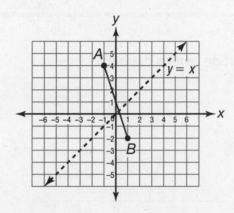

STRATEGY   **Find the new coordinates of the segment.**

STEP 1   Find the new coordinates of $A(-1,4)$. The mapping rule is $(x,y) \rightarrow (y,x)$.

$A(-1,4) \rightarrow A'(4,-1)$

STEP 2   Find the new coordinates of $B(1,-2)$.

$B(1,-2) \rightarrow B'(-2,1)$

SOLUTION   The image of $\overline{AB}$ reflected over the line $y = x$ has endpoints $(4,-1)$ and $(-2,1)$.

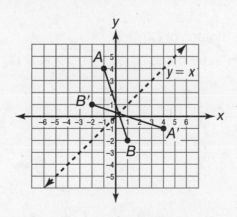

A **point reflection** is a reflection about a fixed point P so that P becomes the midpoint of the segment connecting each point and its image. The notation of a point reflection is $r_{(x,y)}$ where (x,y) is the point of reflection.

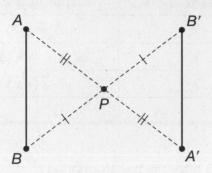

EXAMPLE 3

Draw the triangle after r_o (a reflection across the origin).

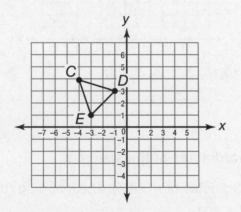

STRATEGY   Draw segments from each point such that (0,0) is the midpoint.

STEP 1   Find the new coordinates of $C(-4,4)$.

Draw $\overline{CO}$, then extend it to C' so that $CO = C'O$.

$C' = (4,-4)$

STEP 2   Find the new coordinates of $D(-1,3)$.

Draw $\overline{DO}$, then extend it to D' so that $DO = D'O$.

$D' = (1,-3)$

STEP 3   Find the new coordinates of $E(-3, 1)$.

Draw $\overline{EO}$, then extend it to E' so that $EO = E'O$.

$E' = (3, -1)$

SOLUTION   **The graph shows the triangle after being reflected over the origin.**

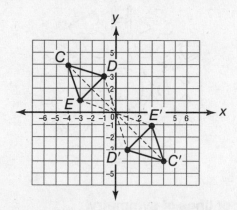

A shape has line symmetry if it can be mapped onto itself with a single reflection. An object has point symmetry if it can be mapped onto itself with a point reflection. Objects with point symmetry also have 180° rotational symmetry.

Line Symmetry   Point Symmetry

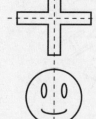
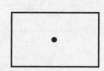

EXAMPLE 4

In a kaleidoscope, mirrors are placed next to each other at congruent angles to create an image with line symmetry. Find the angle at which the mirror should be placed to create the following design.

STRATEGY   **Find the number of lines of symmetry.**

STEP 1   Identify the lines of symmetry in the design.

There are three lines of symmetry.

STEP 2   Find the number of angles in the diagram.

There are 6 angles in the diagram.

STEP 3   Find the measure of each angle.

Because the degrees in a circle are 360° and the 6 angles form a complete circle, divide 360 by 6 to find the measure of a single angle.

$\frac{360}{6} = 60°$

SOLUTION   **The measure of the angle between the mirror in the kaleidoscope should be 60° to create the design above.**

COACHED EXAMPLE

Draw the triangle after $r_{x\text{-axis}}$.

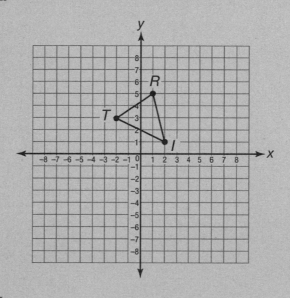

THINKING IT THROUGH

The mapping rule for a figure reflected over the *x*-axis is _____.

The coordinates of $T \rightarrow T'$ are _____ $\rightarrow$ _____.

The coordinates of $R \rightarrow R'$ are _____ $\rightarrow$ _____.

The coordinates of $I \rightarrow I'$ are _____ $\rightarrow$ _____.

Draw and label the preimage, image, and line of reflection in the coordinate plane below.

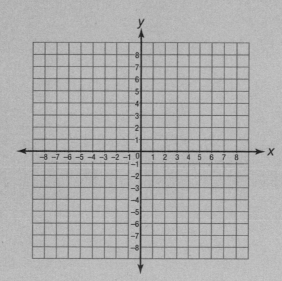

Lesson Practice

Choose the correct answer.

1. Which statement is not true about the relationship between an image and preimage of a reflection?

   (1) They are similar.

   (2) They are congruent.

   (3) They are different sizes.

   (4) They are equidistant from the line of reflection.

2. What are the new coordinates of the point $(-1,2)$ after $r_{x=1}$?

   (1) $(1,2)$

   (2) $(3,2)$

   (3) $(-1,-2)$

   (4) $(2,-1)$

3. How many lines of symmetry does an isosceles triangle have?

   (1) 1

   (2) 2

   (3) 3

   (4) 4

4. Which graph correctly shows the reflection of rectangle $PQRS$ over the line $y = x$?

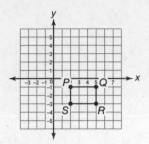

   (1)

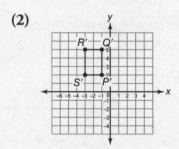

   (2)

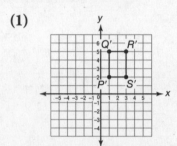

   (3)

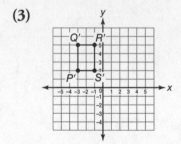

   (4)

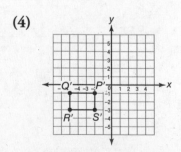

5. Which of the following is the image of the point $(-3,2)$ reflected across the origin?

(1) $(-3,-2)$

(2) $(3,-2)$

(3) $(2,-3)$

(4) $(-2,3)$

6. Triangle *WXY* is reflected across the line $x = -1$. What are the coordinates of the image?

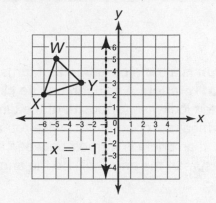

(1) $W'(3,5)$, $X'(4,2)$, $Y'(1,3)$

(2) $W'(-3,5)$, $X'(-4,2)$, $Y'(-1,3)$

(3) $W'(-3,-5)$, $X'(-4,-2)$, $Y'(-1,-3)$

(4) $W'(5,3)$, $X'(2,4)$, $Y'(3,1)$

OPEN-ENDED QUESTION

7. **A.** Write a mapping rule for a reflection across the origin.

_____

B. Use your rule to find the coordinates of the point $(4,-7)$ after r_O.

Rotations

G.G.54, G.G.55, G.G.56, G.G.61

A **rotation** is a transformation that turns a figure about a fixed point. The fixed point is the **center of rotation**. A rotation can be clockwise (in the direction the hands on the clock move) or counterclockwise (opposite of the direction the hands on a clock move). A positive angle of rotation means a counterclockwise rotation. A negative angle of rotation means a clockwise rotation. If an angle is formed between a point, the center of rotation, and the image of the point, the measure of the angle will be the angle of rotation.

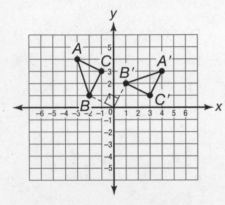

90° Rotation Clockwise

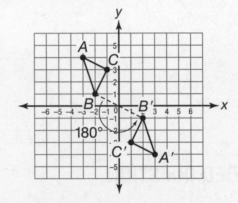

180° Rotation Counterclockwise

A rotation is an isometry. The notation for a rotation is $R_{\text{number of degrees}}$. Mapping rules for common angles of rotation (counterclockwise) are shown in the table.

Angle of Counterclockwise Rotation about the Origin	Mapping Rule	Angle of Clockwise Rotation about the Origin
$R_{90°}$	$P(x,y) \rightarrow P'(-y,x)$	$R_{-270°}$
$R_{180°}$	$P(x,y) \rightarrow P'(-x,-y)$	$R_{-180°}$
$R_{270°}$	$P(x,y) \rightarrow P'(y,-x)$	$R_{-90°}$

EXAMPLE 1

Graph the parallelogram under the rotation $R_{270°}$

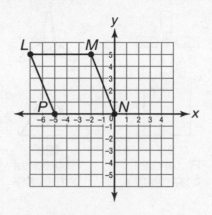

STRATEGY   Use the mapping rules for $R_{270°}$

STEP 1   Find the new coordinates of each point.

The mapping rule is $P(x,y) \rightarrow P'(y,-x)$.

$L(-7,5) \rightarrow L'(5,7)$

$M(-2,5) \rightarrow M'(5,2)$

$N(0,0) \rightarrow N'(0,0)$

$P(-5,0) \rightarrow P'(0,5)$

Notice that the coordinates of a point do not change if it is on the center of rotation like point N.

STEP 2   Graph both figures on the same plane. Check that the angle between a point and its image is 270°.

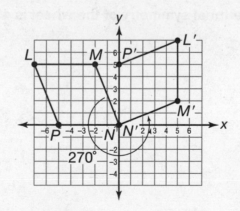

SOLUTION   The graph is shown above.

An object has **rotational symmetry** if it can be mapped onto itself with a rotation of 180° or less.

Objects with Rotational Symmetry

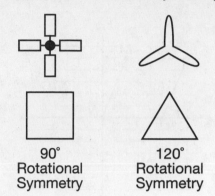

90°
Rotational
Symmetry

120°
Rotational
Symmetry

EXAMPLE 2

Identify the angle of rotational symmetry of the ship's wheel.

STRATEGY   **Divide 360 by the number of congruent angles in the figure.**

There are 8 congruent angles in the diagram, or 4 lines of symmetry.

360 ÷ 8 = 45

SOLUTION   **The angle of rotational symmetry of the wheel is 45°.**

EXAMPLE 3

In the diagram, $\triangle ABC$ is reflected over the *y*-axis. $\triangle A'B'C'$ is then reflected over the *x*-axis. Describe a single transformation that would map $\triangle ABC$ onto $\triangle A''B''C''$.

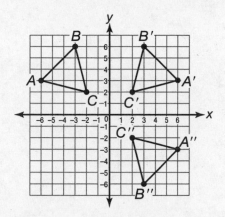

STRATEGY   Examine the graph for an alternate transformation.

  STEP 1   Compare $\triangle ABC$ and $\triangle A''B''C''$.

  $\triangle A''B''C''$ appears to be a 180° rotation of $\triangle ABC$ around the origin.

  STEP 2   Verify your observations with mapping rules.

  The mapping rule for a 180° rotation about the origin is $P(x,y) \rightarrow P'(-x,-y)$.

  $A(-6,3) \rightarrow A''(6,-3)$ These points follow the rule.

  $B(-3,6) \rightarrow B''(3,-6)$ These points follow the rule.

  $C(-2,2) \rightarrow C''(2,-2)$ These points follow the rule.

SOLUTION   The transformation that maps $\triangle ABC$ onto $\triangle A''B''C''$ is $R_{180°}$

COACHED EXAMPLE

Graph figure *WXYZ* after the rotation $R_{-90°}$

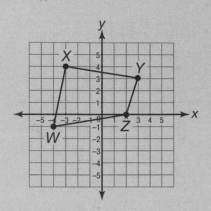

THINKING IT THROUGH

A $-90°$ rotation is the same as a _____ rotation.

The mapping rule for $R_{270°}$ is _____.

Use the rule to find the new coordinates of each point.

$W(\underline{\hspace{1cm}},\underline{\hspace{1cm}}) \rightarrow W'(\underline{\hspace{1cm}},\underline{\hspace{1cm}})$

$X(\underline{\hspace{1cm}},\underline{\hspace{1cm}}) \rightarrow X'(\underline{\hspace{1cm}},\underline{\hspace{1cm}})$

$Y(\underline{\hspace{1cm}},\underline{\hspace{1cm}}) \rightarrow Y'(\underline{\hspace{1cm}},\underline{\hspace{1cm}})$

$Z(\underline{\hspace{1cm}},\underline{\hspace{1cm}}) \rightarrow Z'(\underline{\hspace{1cm}},\underline{\hspace{1cm}})$

Graph the image and preimage in the same plane.

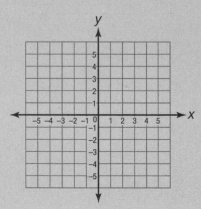

Lesson Practice

Choose the correct answer.

1. Which figure shows the arrow rotated 90° counterclockwise?

   (1)

   (2)

   (3)

   (4)

2. Which of the following figures shows a rotation?

   (1)

   (2)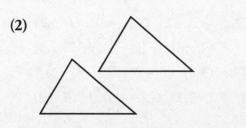

   (3)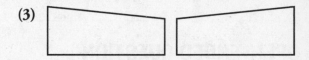

   (4)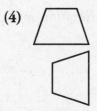

3. Which object has rotational symmetry?

   (1) scalene triangle

   (2) equilateral triangle

   (3) isosceles triangle

   (4) right triangle

4. A triangle has vertices $F(-2,3)$, $G(0,4)$, and $H(2,-3)$. What are the coordinates of $\triangle F'G'H'$ after the rotation $R_{-180°}$?

   (1) $F'(-3,-2)$, $G'(-4,0)$, $H'(3,2)$

   (2) $F'(2,-3)$, $G'(0,-4)$, $H'(-2,3)$

   (3) $F'(3,2)$, $G'(4,0)$, $H'(-3,-2)$

   (4) $F'(-3,2)$, $G'(-4,0)$, $H'(3,-2)$

5. Identify the angle of rotation of the figure below.

   (1) $-90°$

   (2) $-180°$

   (3) $-270°$

   (4) $-360°$

OPEN-ENDED QUESTION

6. Graph the triangle after a rotation $R_{-270°}$

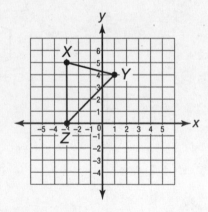

50 Translations

G.G.54, G.G.55, G.G.56, G.G.61

A **translation** moves a figure from one position to another. Each vertex moves the same distance and direction. A translation is an isometry, since it preserves the size of the original. The mapping rule for a translation indicates the number of units that will be added or subtracted from each coordinate, $T_{(a,b)} (x,y) = (x + a, y + b)$.

Translations may also be described using a vector. A **vector** is a ray which indicates both direction and distance. For example, a translation that consists of shifting a figure up two units and left one unit could be written in several ways: $T_{(-1,2)}, (x,y) \rightarrow (x - 1, y + 2)$, or $\vec{v} = <-1,2>$.

EXAMPLE 1

Write the vector in component form.

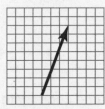

STRATEGY   **Count the units the vector travels.**

   STEP 1   Count the units up.

   The vector rises 8 units.

   STEP 2   Count the units right.

   The vector moves 3 units right.

SOLUTION   **The component form of the vector is $\vec{v} = <3,8>$.**

EXAMPLE 2

Graph the translation of figure *PQRS* under the translation $T_{(6,-4)}$.

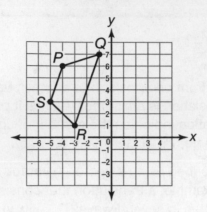

STRATEGY   **Write the transformation as a mapping rule.**

STEP 1   Use the rule to find the new coordinates of each point.

The mapping rule for $T_{(6,-4)}$ is $(x,y) \rightarrow (x + 6, y - 4)$.

$P(-4,6) \rightarrow P'(2,2)$

$Q(-1,7) \rightarrow Q'(5,3)$

$R(-3,1) \rightarrow R'(3,-3)$

$S(-5,3) \rightarrow S'(1,-1)$

STEP 2   Graph both figures on the same coordinate plane.

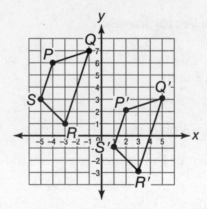

SOLUTION   **The graph is shown above.**

EXAMPLE 3

$\triangle ABC$ is reflected over the line $x = -2$. $\triangle A'B'C'$ is then reflected over the line $x = 3$. Describe the single transformation that maps $\triangle ABC$ onto $\triangle A''B''C''$.

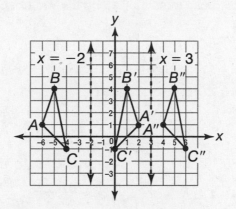

STRATEGY **Write a mapping rule for the transformation.**

STEP 1   Find a mapping rule that describes the transformation of point A to point A''.

The transformation appears to be a horizontal shift.

$A(-6,1) \rightarrow A''(4,1)$

The rule that describes this is $(x,y) \rightarrow (x + 10, y)$.

STEP 2   Verify that the transformation of the other points follow this rule.

$B(-5,4) \rightarrow B''(5,4)$ This point follows the mapping rule.

$C(-4,-1) \rightarrow C''(6,-1)$ This point follows the mapping rule.

SOLUTION   **The transformation that maps $\triangle ABC$ onto $\triangle A''B''C''$ is $T_{(10, 0)}$.**

COACHED EXAMPLE

Graph the figure *KLMN* after it has been translated along the vector shown.

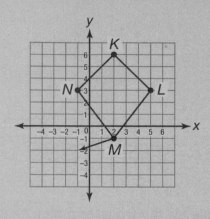

THINKING IT THROUGH

Write the vector in component form.

The vector moves _____ units left and _____ unit down.

The component form of the vector is $\vec{v}$ = <_____,_____>.

Write a mapping rule for the vector.

$(x,y) \rightarrow (x$_____$,y$_____$)$

Use the mapping rule to find the new coordinates of each point.

$K($_____,_____$) \rightarrow K'($_____,_____$)$

$L($_____,_____$) \rightarrow L'($_____,_____$)$

$M($_____,_____$) \rightarrow M'($_____,_____$)$

$N($_____,_____$) \rightarrow N'($_____,_____$)$

Graph the figure *KLMN* after it has been translated along the vector below.

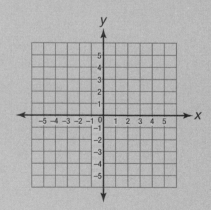

Lesson Practice

Choose the correct answer.

Use the graph for Exercises 1–3.

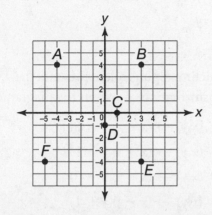

1. Which point is the image of point *A* under $T_{(5,-4)}$?

    (1) *B*

    (2) *C*

    (3) *D*

    (4) *E*

2. Which point is the image of point E under the transformation $(x,y) \rightarrow (x, y + 8)$?

    (1) *A*

    (2) *B*

    (3) *D*

    (4) *F*

3. Which point is the image of point *D* after a translation along $\vec{v} = <-5, -3>$?

    (1) *A*

    (2) *B*

    (3) *E*

    (4) *F*

4. Which graph shows a translation?

    (1)

    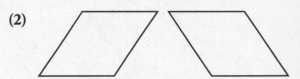

    (2)

    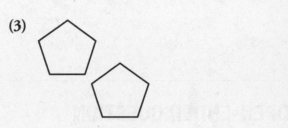

    (3)

    (4)

Use the diagram for Exercises 5 and 6.

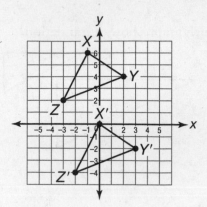

5. Which function rule represents the transformation shown in the diagram?

   (1) $T_{(1,-6)}$

   (2) $T_{(-1,6)}$

   (3) $T_{(-6,1)}$

   (4) $T_{(6,-1)}$

6. Which mapping rule represents the transformation shown in the diagram?

   (1) $(x,y) \rightarrow (x - 1, y + 6)$

   (2) $(x,y) \rightarrow (x + 6, y - 1)$

   (3) $(x,y) \rightarrow (x + 1, y - 6)$

   (4) $(x,y) \rightarrow (x - 6, y + 1)$

OPEN-ENDED QUESTION

7. Let $\overline{A'B'}$ be the image of $\overline{AB}$ after a translation of at least one unit horizontally and one unit vertically. What type of figure is $AA'B'B$? Explain your answer.

_____

_____

_____

Dilations

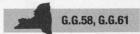

A **dilation** reduces or enlarges a figure. A dilation is not an isometry. The image and preimage are similar but not congruent. A dilation is described using a center of dilation and a scale factor. The **center of dilation** is a fixed point about which all points in the figure are transformed. The **scale factor** is the factor by which the segment is transformed.

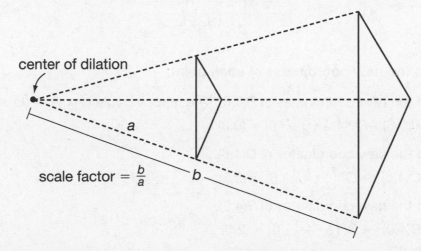

The notation for a dilation is $D_k(x,y)$ where k is the scale factor and (x,y) is the point to be dilated. If $k < 0$ the dilation is also a reflection across the center of dilation. If $0 < k < 1$, the dilation is a reduction. If $k > 0$, the dilation is an enlargement.

> The mapping rule for a dilation with scale factor k centered at the origin is
>
> $$D_k(x,y) \rightarrow (kx, ky)$$

EXAMPLE 1

Draw rectangle *PQRS* after a dilation of $D_{\frac{1}{2}}$ with the center of dilation at the origin.

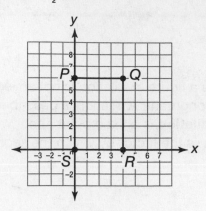

STRATEGY   Find the new coordinates of each point.

STEP 1   Find the new coordinates of *P*(0,6). The mapping rule is $(x,y) \rightarrow (kx,ky)$

$$P(0,6) \rightarrow P'\left(\frac{1}{2} \cdot 0, \frac{1}{2} \cdot 6\right) = (0,3)$$

STEP 2   Find the new coordinates of *Q*(4,6).

$$Q(4,6) \rightarrow Q'\left(\frac{1}{2} \cdot 4, \frac{1}{2} \cdot 6\right) = (2,3)$$

STEP 3   Find the new coordinates of *R*(4,0).

$$R(4,0) \rightarrow R'\left(\frac{1}{2} \cdot 4, \frac{1}{2} \cdot 0\right) = (2,0)$$

STEP 4   Find the new coordinates of *S*(0,0).

$$S(0,0) \rightarrow S'\left(\frac{1}{2} \cdot 0, \frac{1}{2} \cdot 0\right) = (0,0)$$

STEP 5   Draw both rectangles on the same coordinate plane.

SOLUTION   The graph is shown below.

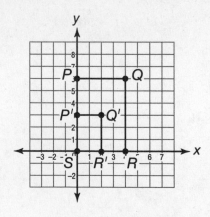

EXAMPLE 2

Tisha wanted to have a 4-inch by 6-inch photo enlarged to be 10 inches long. Find the scale factor for this dilation. Then find the width of the enlarged photo.

STRATEGY   **Find the ratio of new side lengths to old side lengths.**

STEP 1   Find the ratio of lengths.

The old length was 6 inches, and the new length will be 10 inches. Write these as a ratio.

$$\frac{10}{6} = \frac{5}{3}$$

STEP 2   Use the scale factor to find the new width.

$$\frac{5}{3} = \frac{w}{4}$$

$$20 = 3w \quad \text{Cross multiply.}$$

$$6\frac{2}{3} = w$$

SOLUTION   **The scale factor for the enlargement is $\frac{5}{3}$. The new width will be $6\frac{2}{3}$ inches.**

COACHED EXAMPLE

Draw the triangle after the dilation D_{-3}.

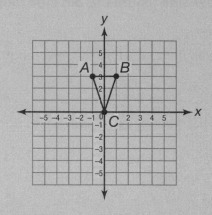

THINKING IT THROUGH

The scale factor of the dilation is negative so the dilation is also _____.

Use the mapping rule to find the new coordinates of each point.

$A(\underline{\hspace{1cm}},\underline{\hspace{1cm}}) \rightarrow A'(\underline{\hspace{0.5cm}}\cdot\underline{\hspace{0.5cm}},\underline{\hspace{0.5cm}}\cdot\underline{\hspace{0.5cm}}) = (\underline{\hspace{1cm}},\underline{\hspace{1cm}})$

$B(\underline{\hspace{1cm}},\underline{\hspace{1cm}}) \rightarrow B'(\underline{\hspace{0.5cm}}\cdot\underline{\hspace{0.5cm}},\underline{\hspace{0.5cm}}\cdot\underline{\hspace{0.5cm}}) = (\underline{\hspace{1cm}},\underline{\hspace{1cm}})$

$C(\underline{\hspace{1cm}},\underline{\hspace{1cm}}) \rightarrow C'(\underline{\hspace{0.5cm}}\cdot\underline{\hspace{0.5cm}},\underline{\hspace{0.5cm}}\cdot\underline{\hspace{0.5cm}}) = (\underline{\hspace{1cm}},\underline{\hspace{1cm}})$

Graph the triangle after the dilation D_{-3}.

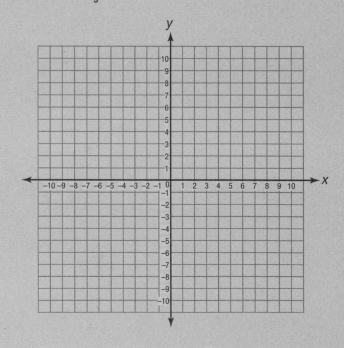

Lesson Practice

Choose the correct answer.

1. Which of the following shows a dilation?

   (1)

   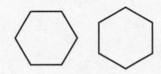

   (2)

   (3)

   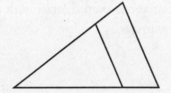

   (4)

   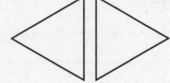

2. Which of the following is not true in a dilation?

   (1) The image and preimage are similar.

   (2) The image and preimage are different shapes.

   (3) The image and preimage have congruent angles.

   (4) The image and preimage have proportional sides.

3. What is the ratio of side length in the preimage to the corresponding side length in the image after the dilation D_4?

   (1) $\frac{1}{4}$

   (2) 0.4

   (3) 4

   (4) 8

4. Which dilation is pictured in the graph below?

   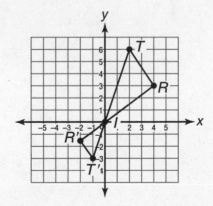

   (1) D_2

   (2) $D_{\frac{1}{2}}$

   (3) D_{-2}

   (4) $D_{\frac{-1}{2}}$

5. Parallelogram *WXYZ* undergoes the dilation $D_{\frac{1}{4}}$. If the side lengths were 8 and 12, what are the side lengths of the image?

   (1) 1, 1.5

   (2) 2, 3

   (3) 4, 8

   (4) 32, 48

6. A square with side length 15 cm is dilated and the image has side length 12 cm. What is the scale factor of the dilation?

   (1) $\frac{4}{5}$

   (2) $\frac{5}{4}$

   (3) 3

   (4) 4

OPEN-ENDED QUESTION

7. A polygon is dilated with a scale factor of $\frac{1}{b}$ and center C. The image is then dilated with a scale factor of b and center C. Describe the size and shape of the new image.

_____

_____

_____

_____

_____

_____

52 Composition of Transformations

G.G.54, G.G.55, G.G.56, G.G.61

A **composition** of transformations occurs when two or more transformations are performed in succession. The composition of two or more isometries is also an isometry. The notation for a composition of transformations is $T_1 \times T_2$. For example, the notation for a composition that consists of a translation followed by a reflection is $r_{y\text{-axis}} \times T_{(1,2)}$. Notice that the transformation listed first is performed after the transformation listed second. This is because the order of performing transformations is from right to left.

EXAMPLE 1

Draw the triangle after the composition $R_{90} \times T_{(0,-6)}$.

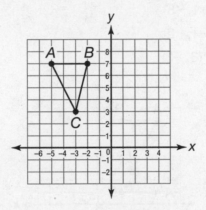

STRATEGY   Apply mapping rules for one transformation at a time.

STEP 1   Apply the second transformation first. The translation rule is $(x,y) \rightarrow (x, y - 6)$.

$$A(-5,7) \rightarrow A'(-5, 7 - 6) = (-5,1)$$
$$B(-2,7) \rightarrow B'(-2, 7 - 6) = (-2,1)$$
$$C(-3,3) \rightarrow C'(-3, 3 - 6) = (-3,-3)$$

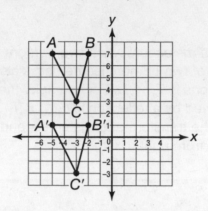

STEP 2   Apply the rotation. The mapping rule is $P(x,y) \rightarrow P'(-y,x)$.

$$A'(-5,1) \rightarrow A''(-1,-5)$$
$$B'(-2,1) \rightarrow B''(-1,-2)$$
$$C'(-3,-3) \rightarrow C''(3,-3)$$

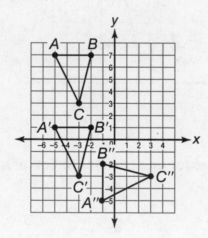

SOLUTION   **The graph is shown above.**

When a translation is followed by a reflection, the result is a **glide reflection**. The vector of translation and the line of reflection must be parallel and the order of transformations does not matter. Because translations and reflections are both isometries, a glide reflection is also an isometry.

EXAMPLE 2

Draw the figure under the glide reflection $r_{y=x} \times T_{(4,4)}$

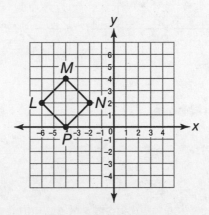

STRATEGY   **Apply mapping rules for one transformation at a time.**

STEP 1   Apply the translation. The mapping rule is $(x,y) \rightarrow (x + 4, y + 4)$

$L(-6,2) \rightarrow L'(-6 + 4, 2 + 4) = (-2,6)$

$M(-4,4) \rightarrow M'(-4 + 4, 4 + 4) = (0,8)$

$N(-2,2) \rightarrow N'(-2 + 4, 2 + 4) = (2,6)$

$P(-4,0) \rightarrow P'(-4 + 4, 0 + 4) = (0,4)$

STEP 2   Apply the reflection. The mapping rule is $(x,y) \rightarrow (y,x)$.

$L'(-2,6) \rightarrow L''(6,-2)$

$M'(0,8) \rightarrow M''(8,0)$

$N'(2,6) \rightarrow N''(6,2)$

$P'(0,4) \rightarrow P''(4,0)$

STEP 3   Draw all three figures on the same coordinate plane.

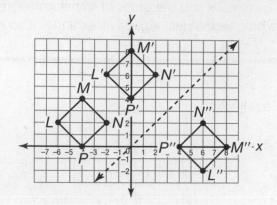

SOLUTION   The graph is shown above.

COACHED EXAMPLE

Write a mapping rule for the composition that consists of a dilation centered at the origin with scale factor of $\frac{3}{2}$, then a translation along the vector $\langle -2,4 \rangle$.

THINKING IT THROUGH

Write the composition in proper notation.

T_____ $\times D$_____

Write the rule by applying one transformation at a time.

For the dilation with scale factor $\frac{3}{2}$, the mapping rule is $(x,y) \rightarrow ($_____,_____$)$.

Add the mapping rule for the translation.

$(x,y) \rightarrow \left(\frac{3}{2}x \underline{\hspace{2cm}}, \frac{3}{2}y \underline{\hspace{2cm}} \right)$

The mapping rule is T_____ $\times D$_____

$(x,y) \rightarrow ($_____,_____$)$.

Lesson Practice

Choose the correct answer.

1. Which of the following shows a composition?

   (1)

   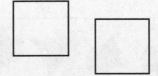

   (2)

   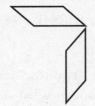

   (3)

   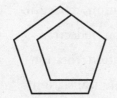

   (4)

   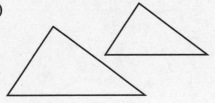

2. Which of the following shows a glide reflection?

   (1)

   (2)

   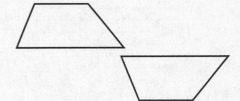

   (3)

   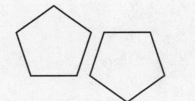

   (4)

3. What composition is shown in the graph below?

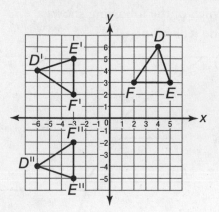

(1) $R_{90°} \times r_{x\text{-axis}}$

(2) $r_{x\text{-axis}} \times R_{90°}$

(3) $R_{90°} \times T_{(0,-7)}$

(4) $T_{(0,-7)} \times R_{-90°}$

4. Which of the following is not true about a glide reflection?

(1) It is an isometry.

(2) The line of reflection is perpendicular to the translation vector.

(3) It is a composition of a translation and a reflection.

(4) It is commutative.

5. What are the coordinates of the image of $\triangle G(0,-1) H(2,-5) I(-2,-5)$ after the composition $T_{(-1,2)} \times R_{180°}$?

(1) $G''(1,-1)$, $H''(-1,3)$, $I''(3,3)$

(2) $G''(-1,3)$, $H''(-3,7)$, $I''(1,7)$

(3) $G''(-2,2)$, $H''(-4,6)$, $I''(0,6)$

(4) $G''(2,0)$, $H''(0,4)$, $I''(4,4)$

6. Gina is using a two-rose stencil to create the pattern shown. What composition of transformations should she use to create the pattern?

(1) translation, reflection

(2) rotation, reflection

(3) translation, rotation

(4) dilation, translation

OPEN-ENDED QUESTION

7. Does the following diagram show a glide reflection? Why or why not?

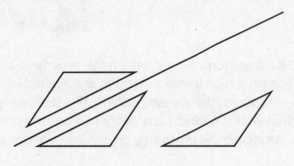

_____

_____

_____

_____

_____

53 Properties of Transformations

 G.G. 54, G.G. 55, G.G. 56, G.G. 57, G.G. 59, G.G. 60

Recall that an isometry is a transformation in which the preimage and image are congruent. There are 6 characteristics on which transformations are evaluated. The characteristics are distance (length), angle measures, parallelism, collinearity, betweeness, and orientation (clockwise vs. counterclockwise). A **direct isometry** is a transformation that preserves all six of these characteristics. An **opposite isometry** preserves everything but orientation.

EXAMPLE 1

$\triangle A'B'C'$ is the image of $\triangle ABC$ after a 90° rotation about the origin. Determine if a rotation is a direct isometry.

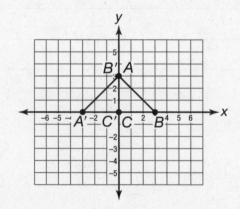

STRATEGY **Verify each property of a direct isometry using coordinate geometry.**

STEP 1 Determine if the distances have changed using the distance formula.

$$d = \sqrt{(x_2 - x_1)^2 + (y_2 - y_1)^2}$$

$$AB = \sqrt{(0 - 3)^2 + (3 - 0)^2} = \sqrt{9 + 9} = \sqrt{18} = 3\sqrt{2}$$

$$A'B' = \sqrt{(-3 - 0)^2 + (0 - 3)^2} = \sqrt{9 + 9} = \sqrt{18} = 3\sqrt{2}$$

$$AC = 3$$

$$A'C' = 3$$

$$BC = 3$$

$$B'C' = 3$$

Distance is preserved.

STEP 2   Determine if angle measures are preserved.

$\triangle ABC \cong \triangle A'B'C'$ (using SSS), therefore corresponding angles are congruent.

Angle measure is preserved.

STEP 3   Determine if parallelism is preserved.

Because the image and preimage are congruent, parallelism is preserved.

STEP 4   Determine if collinearity is preserved.

A is collinear with B and C, A' is collinear with B' and C'.

Collinearity is preserved.

STEP 5   Determine if betweenness is preserved.

Let point P be on $\overline{AB}$, then after the rotation, point P' would be on $\overline{A'B'}$.

Betweenness is preserved.

STEP 6   Determine if orientation is preserved.

In $\triangle ABC$, the vertices are labeled in a clockwise order. Similarly, in $\triangle A'B'C'$, vertices are labeled in a clockwise order.

SOLUTION   **Because it preserves all six characteristics, rotation is a direct isometry.**

The table shows which transformations preserve which characteristics.

Property	Line Reflection	Point Reflection	Translation	Rotation	Dilation	Glide Reflection
Distance	X	X	X	X		X
Angle Measure	X	X	X	X	X	X
Parallelism	X	X	X	X	X	X
Collinearity	X	X	X	X	X	X
Betweenness	X	X	X	X	X	X
Orientation		X	X	X	X	
Isometry	Opposite	Direct	Direct	Direct	No	Opposite

EXAMPLE 2

In the diagram, $\triangle L'M'N$ is the image of $\triangle LMN$ after a reflection over point N.

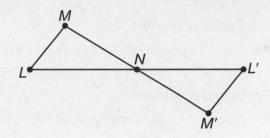

Show that $\overline{LM} \parallel \overline{L'M'}$.

STRATEGY   **Use properties of Direct Isometry.**

STEP 1   Show that alternate interior angles are congruent.

Because a direct isometry preserves angle measures, $\angle L \cong \angle L'$.

STEP 2   Show that the sides are parallel.

Because alternate interior angles are congruent, $\overline{LM} \parallel \overline{L'M'}$.

SOLUTION   **The steps above show $\overline{LM} \parallel \overline{L'M'}$.**

EXAMPLE 3

The figure below △W'XY' is the image of △WXY under some isometry. Identify the isometry and justify your assertion using properties of transformations.

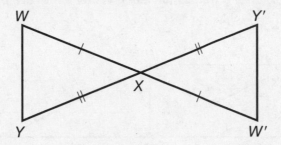

STRATEGY   **Determine which properties the transformation preserves.**

STEP 1   Show that △W'XY' ≅ △WXY.

It is given that $\overline{WX} \cong \overline{W'X}$ and $\overline{XY} \cong \overline{XY'}$.

∠WXY ≅ ∠W'XY' because they are vertical angles.

Therefore, △WXY ≅ △W'XY' using SAS.

STEP 2   Identify the transformation.

The transformation appears to be a reflection, however, orientation is preserved since both triangles are labeled in a clockwise fashion. Therefore, the transformation is a point reflection or 180° rotation.

SOLUTION   **Since the transformation that maps △WXY onto △W'XY' is either a point reflection or a 180° rotation, the isometry is direct.**

COACHED EXAMPLE

$\triangle P'RR'$ is the image of $\triangle PQR$ after a translation along vector QR. Prove that $PRR'P'$ is a parallelogram.

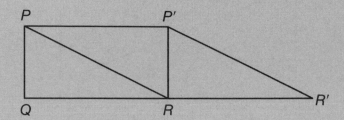

THINKING IT THROUGH

A translation is an isometry so $\triangle PQR$ _____ $\triangle P'RR'$.

Corresponding angles, $\angle$_____ $\cong$ $\angle$_____ by CPCTC.

Therefore, _____ $\parallel$ _____.

Also by CPCTC, $\overline{PR} \cong$ _____.

So, $PRR'P'$ is a parallelogram because _____

_____

_____ .

Lesson Practice

Choose the correct answer.

1. Which of the following transformations is an opposite isometry?

   (1) rotation
   (2) line reflection
   (3) translation
   (4) dilation

2. An opposite isometry preserves all of the following properties except

   (1) distance.
   (2) parallelism.
   (3) betweenness.
   (4) orientation.

3. A dilation preserves all of the following properties except

   (1) distance.
   (2) parallelism.
   (3) betweenness.
   (4) orientation.

4. What transformation will map $\triangle ADF$ onto $\triangle AD'F$?

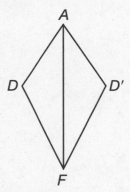

   (1) rotation
   (2) point reflection
   (3) line reflection
   (4) translation

5. In the diagram, $\overline{C'D'}$ is the image of $\overline{CD}$ after a reflection across the y-axis. What type of quadrilateral is $CDD'C'$?

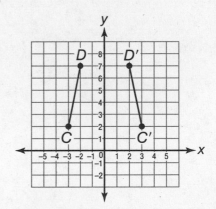

   (1) rectangle
   (2) parallelogram
   (3) isosceles trapezoid
   (4) square

OPEN-ENDED QUESTION

6. Graph $\overline{AB}$ where $A(0,3)$ and $B(3,0)$. Let $\overline{A'B'}$ be the image of $\overline{AB}$ under a point reflection in the origin. What type of quadrilateral is $ABA'B'$? Prove your assertion.

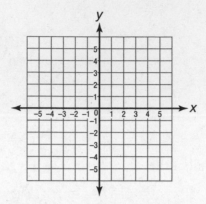

_____

_____

_____

_____

_____

New York State Regents Examination Coach, Geometry

PRACTICE TEST 1

Name: _____

Part I

Answer all questions in this part. Each correct answer will receive 2 credits. No partial credit will be allowed. For each question, write on the separate answer sheet the numeral preceding the word or expression that best completes the statement or answers the question.

1. Which graph matches the equation $(x - 1)^2 + (y - 3)^2 = 9$?

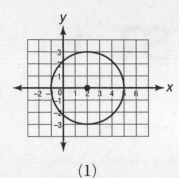

(1)

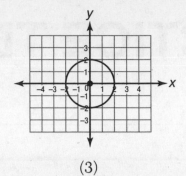

(3)

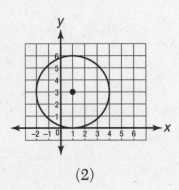

(2)

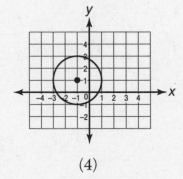

(4)

2. The tangent to circle O is $\overline{PQ}$. The base of $\triangle PQO$, $\overline{PO}$ is the radius of circle O. If the radius equals 9 in. and the tangent equals 15 in., what is the length of $\overline{OQ}$? Round to nearest whole number.

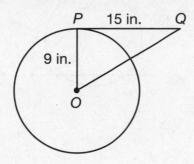

(1) 8 in.

(2) 12 in.

(3) 17 in.

(4) 23 in.

3. Intersecting lines a and b are in plane R. Line m is perpendicular to both lines a and b. Line m also satisfies which following condition?

(1) Line m is parallel to lines a and b.

(2) Line m is skew to lines a and b.

(3) Line m is perpendicular to plane R.

(4) Line m is parallel to plane R.

4. What is the inverse of the statement "If an angle is a straight angle, then its measure is 180°"?

(1) If the measure of an angle is 180°, then it is a straight angle.

(2) If an angle is not a straight angle, then its measure is not 180°.

(3) If the measure of an angle is not 180°, then it is not a straight angle.

(4) An angle is a straight angle if and only if its measure is 180°.

PRACTICE TEST 1

5. Find the value of x.

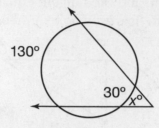

(1) 30°   (3) 50°

(2) 40°   (4) 60°

6. What is the measure of each exterior angle of a regular octagon?

(1) 45°   (3) 135°

(2) 90°   (4) 180°

7. Find the approximate volume of a square pyramid with a base edge of 7 cm and an altitude of 16 cm. Round to the nearest cubic centimeter.

(1) 112 cm^3 (3) 512 cm^3

(2) 261 cm^3 (4) 784 cm^3

8. Find the coordinates of the midpoint of the line segment with coordinates $(3, -5)$ and $(-7, -11)$.

(1) $(-2, 8)$ (3) $(2, 8)$

(2) $(2, -8)$ (4) $(-2, -8)$

9. Point E is on line a. How many planes are perpendicular to line a through point E?

    (1) none

    (2) 1

    (3) 2

    (4) infinite

10. Given parallelogram $ABCD$ with $\triangle ABC \cong \triangle ADC$. $\angle D \cong \angle B$ and $\angle DAC \cong \angle BCA$. Which statement shows corresponding parts that are congruent?

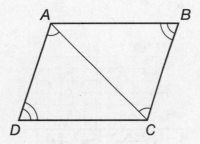

    (1) $\angle DAC \cong \angle B$

    (2) $\angle BCA \cong \angle D$

    (3) $\overline{AB} \cong \overline{BC}$

    (4) $\overline{BA} \cong \overline{DC}$

11. $\overleftrightarrow{CG}$ is perpendicular to plane X at point G. $\overleftrightarrow{DE}$ is perpendicular to plane X at point E. $\overleftrightarrow{CG}$ and $\overleftrightarrow{DE}$ are

    (1) collinear.

    (2) intersecting.

    (3) parallel.

    (4) skew.

12. Given $\triangle PNB$ has coordinates $P(2,2)$, $N(3,-1)$, and $B(-1,-2)$, what is the image under the glide reflection $<2,0>$ and $y = 3$?

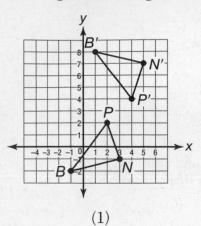

(1)

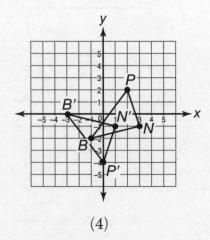

(3)

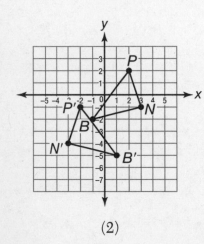

(2)

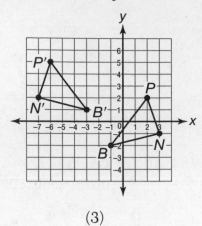

(4)

13. Describe the loci of points A, B, C, and D shown in the diagram.

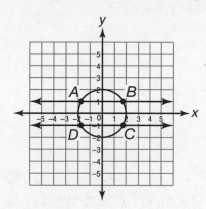

(1) 2 units from the origin and 1 unit from the y-axis

(2) 1 unit from the origin and 2 units from the y-axis

(3) 2 units from the origin and 1 unit from the x-axis

(4) 1 unit from the origin and 2 units from the x-axis

14. The rectangular recreation area for the paddle-boats at a park has a length of 350 ft and a width of 250 ft. The two lake docks are at diagonal corners. What is the distance between the docks to the nearest foot?

(1) 430 ft

(2) 600 ft

(3) 875 ft

(4) 1200 ft

15. Identify the center and radius of the circle with the equation $(x + 9)^2 + (y - 4)^2 = 5$.

(1) center $(9, -4)$; radius $= 5$

(2) center $(-9, -4)$; radius $= \sqrt{5}$

(3) center $(-9, 4)$; radius $= 5$

(4) center $(-9, 4)$; radius $= \sqrt{5}$

16. $\overleftrightarrow{EF}$ is perpendicular to plane G. How many other lines are perpendicular to plane G through point F?

(1) infinite

(2) 0

(3) 1

(4) 2

17. In $\triangle ABC$, side $\overline{AC} = 10$ in., $\overline{BC} = 14$ in., and $\overline{AB} = 8$ in. The midpoint of $\overline{AB}$ is D and the midpoint of $\overline{AC}$ is E. What is the length of $\overline{DE}$?

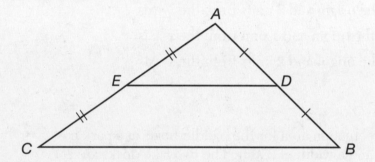

(1) 4 in.

(2) 5 in.

(3) 7 in.

(4) 8 in.

18. The graph of the linear equations: $y = 2x - 7$ and $y = -2x + 7$ shows

(1) intersecting lines.

(2) parallel lines.

(3) perpendicular lines.

(4) skew lines.

19. In a coordinate plane, $\triangle GPS$ has coordinates $G(2,1)$, $P(4,0)$, and $S(5,7)$. $\triangle GPS$ can be proven to be a right triangle using only

(1) the Distance Formula.

(2) the Midpoint Formula.

(3) the Slope Formula.

(4) the Transitive Property.

20. Two parallel planes E and F are intersected by plane G. What figure is formed by the intersection?

(1) one parallel plane and one line

(2) one parallel plane and two intersecting lines

(3) two parallel planes and one line

(4) two parallel lines

21. What is the measure of $\angle x$?

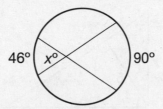

(1) 46°

(2) 68°

(3) 90°

(4) 136°

22. What is the image of point $T(3,4)$ after a 90° counterclockwise rotation about the origin?

(1) $(4,-3)$

(2) $(-4,3)$

(3) $(4,3)$

(4) $(-4,-3)$

23. Points E, G, and H are collinear. In $\triangle EFG$, $\angle E = 53°$ and $\angle F = 67°$. Exterior $\angle FGH$ must be

(1) 60°.

(2) 120°.

(3) 150°.

(4) 180°.

24. In the right triangle below, $y = 2\sqrt{13}$. Find x.

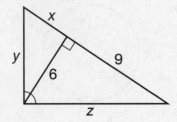

(1) 3

(2) $3\sqrt{5}$

(3) $3\sqrt{13}$

(4) 4

25. To begin the construction of equilateral $\triangle ABC$ given $\overline{AB}$, open the compass to the length of $\overline{AB}$. Place the compass on point A. Draw an arc. Keep the same length on the compass and draw an arc that intersects the first arc from

   (1) B.

   (2) C.

   (3) $\overline{BC}$.

   (4) $\overline{AC}$.

26. Find the lateral area of a right circular cylinder if the diameter is 8 in. and the altitude is 6 in.

   (1) 24π in.^{2}

   (2) 40π in.^{2}

   (3) 48π in.^{2}

   (4) 80π in.^{2}

27. The opposite lateral edges of an octagonal prism are

   (1) different lengths.

   (2) perpendicular.

   (3) congruent and perpendicular.

   (4) congruent and parallel.

28. $\overline{AB}$ is in plane N. Plane N is perpendicular to plane M if and only if

   (1) $\overline{AB} \perp$ plane N.

   (2) plane M contains $\overline{AB}$.

   (3) $\overline{AB} \perp$ plane M or is the line of intersection of plane N and plane M.

   (4) plane N contains $\overline{AB}$.

Part II

Answer all the questions in this part. Each correct answer will receive 2 credits. Clearly indicate the necessary steps, including appropriate formula substitutions, diagrams, graphs, charts, etc. For all questions in this part, a correct numerical answer with no work shown will receive only 1 credit.

29. What type of transformation maps $QRST$ onto $Q'R'S'T'$?

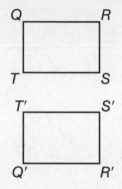

30. Construct the bisector of the angle using a compass and straightedge. Be sure to show all construction marks.

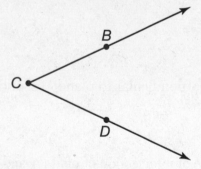

31. In the diagram, $\overleftrightarrow{LM} \parallel \overleftrightarrow{PN}$. What is the measure of $\overarc{PN}$?

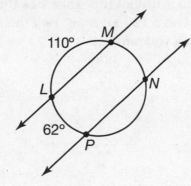

32. $\triangle LNO$ is the image of $\triangle LMP$ after $D_{\frac{4}{3}}$. Find $\overline{LN}$ and $\overline{LO}$.

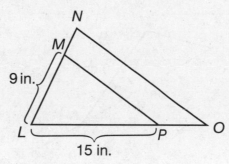

33. Mrs. Thomas has two dogs, Rusty and Randy. She ties Randy to a post in the middle of her yard on a 10-foot rope. She ties Rusty to a post 20 feet from Randy's post on a 12-foot rope. Will the dogs be able to reach each other? Explain your answer.

34. Show that △*LMN* ~ △*OPN*.

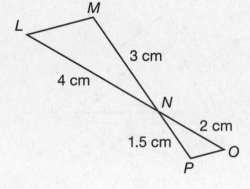

Duplicating any part of this book is prohibited by law.

Part III

Answer all the questions in this part. Each correct answer will receive 3 credits. Clearly indicate the necessary steps, including appropriate formula substitutions, diagrams, graphs, charts, etc. For all questions in this part, a correct numerical answer with no work shown will receive only 1 credit.

35. Find the equation of the line perpendicular to $y = 3x - 1$, through the point (6,3).

PRACTICE TEST 1

36. Solve the system of equations by graphing.

$$y = x^2 - 4x - 2$$
$$y = x - 2$$

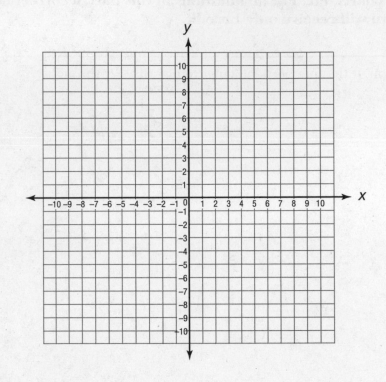

37. Izekial wants to send a baseball bat that is 34 inches long to his cousin. The post office has square boxes with a height of 5 inches. What length and width to the nearest inch, is the smallest square box in which Izekial can send the bat if he lays it flat on the bottom of the box and across the diagonal?

Part IV

Answer all the questions in this part. Each correct answer will receive 4 credits. Clearly indicate the necessary steps, including appropriate formula substitutions, diagrams, graphs, charts, etc. For all questions in this part, a correct numerical answer with no work shown will receive only 1 credit.

38. Given: $\triangle HEG \cong \triangle FGE$
Prove: $EFGH$ is a parallelogram.

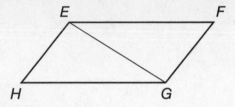

New York State Regents
Examination Coach, Geometry

PRACTICE TEST 2

Name: _____

Part I

Answer all questions in this part. Each correct answer will receive 2 credits. No partial credit will be allowed. For each question, write on the separate answer sheet the numeral preceding the word or expression that best completes the statement or answers the question.

1. Isosceles trapezoid $ABCD$ has diagonals $\overline{AC} = 4x$ and $\overline{BD} = 3y$. The legs are $\overline{AB} = 3x - 5$ and $\overline{DC} = x + y$. Find the length of diagonal $\overline{AC}$.

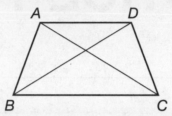

   (1) 7.5   (3) 17.5

   (2) 10   (4) 30

2. Find the equation of the perpendicular bisector of the line segment that has endpoints $(3, -6)$ and $(-5, 4)$.

   (1) $y = \frac{4}{5}x + \frac{1}{5}$ (3) $y = -\frac{5}{4}x - 1$

   (2) $y = \frac{4}{5}x - \frac{1}{5}$ (4) $y = -\frac{5}{4}x + 1$

3. Isosceles $\triangle ABC$ has a base angle of $44°$. What is the measure of the vertex angle?

   (1) $22°$ (3) $88°$

   (2) $44°$ (4) $92°$

4. $\triangle FGH$ contains $\overline{FG} \parallel \overline{IJ}$. $IJ = 7$ cm, $FG = 14$ cm, $HJ = w$, and $JG = 9.2$ cm. Find the value of w.

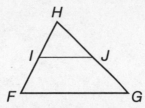

    (1) 3.5 cm

    (2) 4.6 cm

    (3) 7 cm

    (4) 9.2 cm

5. What is the equation of the circle in the graph below?

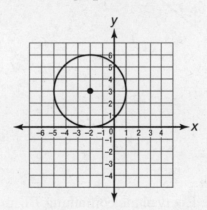

    (1) $(x + 2)^2 + (y - 3)^2 = 9$

    (2) $(x + 3)^2 + y^2 = 9$

    (3) $(x - 2)^2 + (y - 2)^2 = 9$

    (4) $(x + 1)^2 + (y + 2)^2 = 9$

6. Find the slope of the line perpendicular to the line $y = -3x - 5$.

(1) -3

(2) $-\frac{1}{3}$

(3) $\frac{1}{3}$

(4) 3

7. What postulate or theorem proves the triangles below are similar?

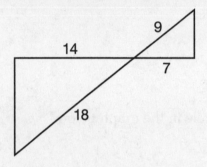

(1) AA

(2) ASA

(3) SAS

(4) SSS

8. $\overleftrightarrow{PT}$ is perpendicular to plane X. Every plane containing $\overleftrightarrow{PT}$ must be

(1) parallel to plane X.

(2) skew to plane X.

(3) perpendicular to plane X.

(4) in plane X.

9. Quadrilateral *ABCD* is a parallelogram because

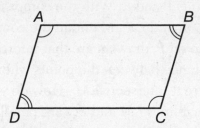

(1) two of the angles have equal measures.

(2) two of the opposite sides are congruent.

(3) two of the opposite sides are parallel.

(4) two pairs of opposite angles are equal.

10. Planes *J* and *K* are each perpendicular to $\overleftrightarrow{SW}$. Planes *J* and *K* must be

(1) intersecting.

(2) parallel.

(3) perpendicular.

(4) skew.

11. To construct a line parallel to $\overleftrightarrow{LM}$ through a given point P not on the line, draw a line connecting points P and L. With the compass tip on L, draw an arc that intersects $\overleftrightarrow{LM}$ and $\overleftrightarrow{LP}$. Maintain the compass setting and with the compass tip on P, draw an arc that intersects $\overleftrightarrow{LP}$. Set the compass to the distance between the points where the first arc intersects $\overleftrightarrow{LP}$ and where it intersects $\overleftrightarrow{LM}$. How do you complete the construction so that lines LM and PQ are parallel?

(1) Place your compass tip at P and draw an arc that intersects $\overleftrightarrow{LP}$. Label the point of intersection Q. Draw $\overleftrightarrow{PQ}$.

(2) Place your compass tip where the first arc intersects $\overleftrightarrow{LM}$ and draw an arc that does intersects $\overleftrightarrow{LM}$. Label the point of intersection Q. Draw $\overleftrightarrow{PQ}$.

(3) Place your compass tip where the second arc intersects $\overleftrightarrow{LP}$ on the outside of point P. Draw an arc that intersects the second arc. Label the point of intersection Q. Draw $\overleftrightarrow{PQ}$.

(4) Place your compass tip where the second arc intersects $\overleftrightarrow{LP}$. Draw an arc that intersects $\overleftrightarrow{LP}$. Label the point of intersection Q. Draw $\overleftrightarrow{PQ}$.

12. What is the truth value for the negation of the statement below?

"A right angle measures 90°."

(1) always true

(2) always false

(3) sometimes true

(4) sometimes false

13. Two sides of a triangle are 9 cm and 17 cm. Which of the following cannot be the length of the third side?

(1) 8 cm

(2) 10 cm

(3) 14 cm

(4) 16 cm

14. The equation of the circle with diameter endpoints of $(1,1)$ and $(-9,1)$ is

  (1) $(x - 4)^2 + (y + 1)^2 = 5$.

  (2) $(x + 4)^2 + (y + 1)^2 = 5$.

  (3) $(x - 4)^2 + (y - 1)^2 = 25$.

  (4) $(x + 4)^2 + (y - 1)^2 = 25$.

15. $\overleftrightarrow{GT}$ is perpendicular to plane U at point T. $\overrightarrow{ET}$ is perpendicular to $\overleftrightarrow{GT}$ at T. Which statement is true about $\overrightarrow{ET}$?

  (1) It is parallel to plane U.

  (2) It is contained in plane U.

  (3) It is perpendicular to plane U.

  (4) It is the same line as $\overleftrightarrow{GT}$.

16. Two lines are cut by a transversal. The same side interior angles are 108° and 72°. The lines must be

  (1) intersecting.

  (2) parallel.

  (3) skew.

  (4) supplementary.

17. What theorem or postulate proves the triangles congruent?

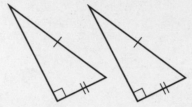

(1) SSS

(2) SAS

(3) ASA

(4) HL

18. The diagonals of which quadrilateral are always congruent?

(1) rectangle

(2) rhombus

(3) trapezoid

(4) parallelogram

19. $ABCD$ with $A(2,2)$, $B(-2,2)$, $C(-2,-2)$, and $D(2,-2)$ is transformed by $D_{-3}(0,0)$. What conclusion can be drawn from the effect of this transformation?

(1) The image is an enlargement rotated $180°$ about the origin.

(2) The image is a reduction rotated $90°$ about the origin.

(3) The image is a reduction rotated $270°$ about the origin.

(4) The image is an enlargement rotated $-90°$ about the origin.

20. $\triangle HKM$ has $\angle K = 62°$ and $\angle H = 84°$. What is the measure of $\angle M$?

   (1) 34°

   (2) 44°

   (3) 90°

   (4) 180°

21. The intersection of a plane with a sphere is

   (1) an ellipse.

   (2) a circle.

   (3) a parabola.

   (4) a hyperbola.

22. What is the sum of the exterior angles of any polygon?

   (1) 90°

   (2) 180°

   (3) 270°

   (4) 360°

23. In a regular hexagonal pyramid, all lateral faces are congruent

   (1) isosceles triangles.

   (2) rectangles.

   (3) hexagons

   (4) right isosceles triangles.

24. The opposite angles in which polygon must be congruent?

    (1) quadrilateral

    (2) pentagon

    (3) parallelogram

    (4) trapezoid

25. Find the equation of the line containing point (9,2) parallel to line
$y = -x + 7$.

    (1) $y = -x + 11$

    (2) $y = x - 9$

    (3) $y = -x - 11$

    (4) $y = x + 9$

26. The centroid of $\triangle ABC$ is point P. N is the midpoint of $\overline{AC}$. What is
the value of the ratio $BP{:}PN$?

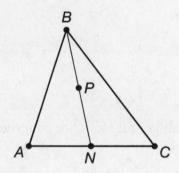

    (1) 1:2

    (2) 1:3

    (3) 3:1

    (4) 2:1

27. Find the lateral area of a circular cone given the diameter is 6 ft and the slant height is 8 ft.

(1) 12π ft^2

(2) 24π ft^2

(3) 36π ft^2

(4) 48π ft^2

28. What figure would this construction produce?

Set the compass so that it is wider than one-half the length of the given $\overline{AB}$. Put the compass point on point A and draw an arc above and below $\overline{AB}$. Keep the same compass setting and draw an arc from point B above and below $\overline{AB}$ intersecting the first arc above and below $\overline{AB}$. Label the points of intersection Y and Z respectively. Draw $\overline{YZ}$ through the intersection points.

(1) the angle bisector

(2) parallel lines

(3) the perpendicular bisector

(4) a copied angle

PRACTICE TEST 2

Part II

Answer all the questions in this part. Each correct answer will receive 2 credits. Clearly indicate the necessary steps, including appropriate formula substitutions, diagrams, graphs, charts, etc. For all questions in this part, a correct numerical answer with no work shown will receive only 1 credit.

29. In the diagram, point E is the midpoint of $\overline{AD}$ and point B is the midpoint of $\overline{AC}$. Identify the dilation that maps $\triangle AEB$ onto $\triangle ADC$.

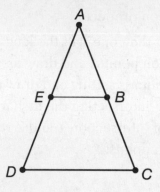

30. What are the new coordinates of the end points of $\overline{GH}$ with $G(5,2)$ and $H(-3, -1)$ after a $T_{(-2,4)}$?

31. Two prisms have the bases shown in the diagram and heights h. How are the volumes of the prisms related? Support your answer with calculations.

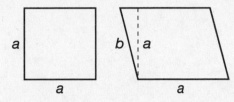

32. In $\triangle XYZ$, $XY = 7$ in., $YZ = 9$ in., and $XZ = 4$ in. Name the smallest and largest angle in the triangle.

33. Construct a truth table for the statement $\sim(p \vee q) \rightarrow \sim p \vee \sim q$.

34. $\overline{OP}$ is a radius of circle O. $AB = 6$ inches. Find AM.

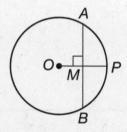

Part III

Answer all questions in this part. Each correct answer will receive 3 credits. Clearly indicate the necessary steps, including appropriate formula substitutions, diagrams, graphs, charts, etc. For all questions in this part, a correct numerical answer with no work shown will receive only 1 credit.

35. Find the length of the line segment with endpoints $(1, -1)$ and $(4,5)$.

36. Given that point P is the centroid, find the perimeter of the triangle.

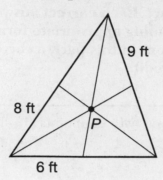

9 ft

8 ft

P

6 ft

37. What is the measure of each interior angle in a regular octagon?

Part IV

Answer all questions in this part. Each correct answer will receive 4 credits. Clearly indicate the necessary steps, including appropriate formula substitutions, diagrams, graphs, etc. For all questions in this part, a correct numerical answer with no work shown will receive only 1 credit.

38. Graph $\triangle QRS$ after $r_{x\text{-axis}}$. Use properties of transformations to determine what type of quadrilateral is formed by $QRSR'$.

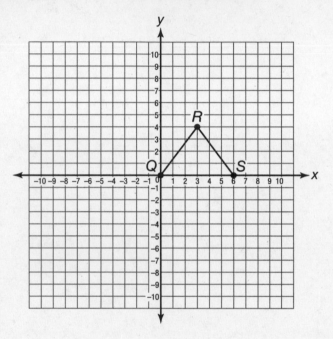

Geometry Reference Sheet

Volume	Cylinder	$V = Bh$ where B is the area of the base
	Pyramid	$V = \frac{1}{3}Bh$ where B is the area of the base
	Right Circular Cone	$V = \frac{1}{3}Bh$ where B is the area of the base
	Sphere	$V = \frac{4}{3}\pi r^3$

Lateral Area (L)	Right Circular Cylinder	$L = 2\pi rh$
	Right Circular Cone	$L = \pi rl$ where l is the slant height

Surface Area	Sphere	$SA = 4\pi r^2$

Graph Paper

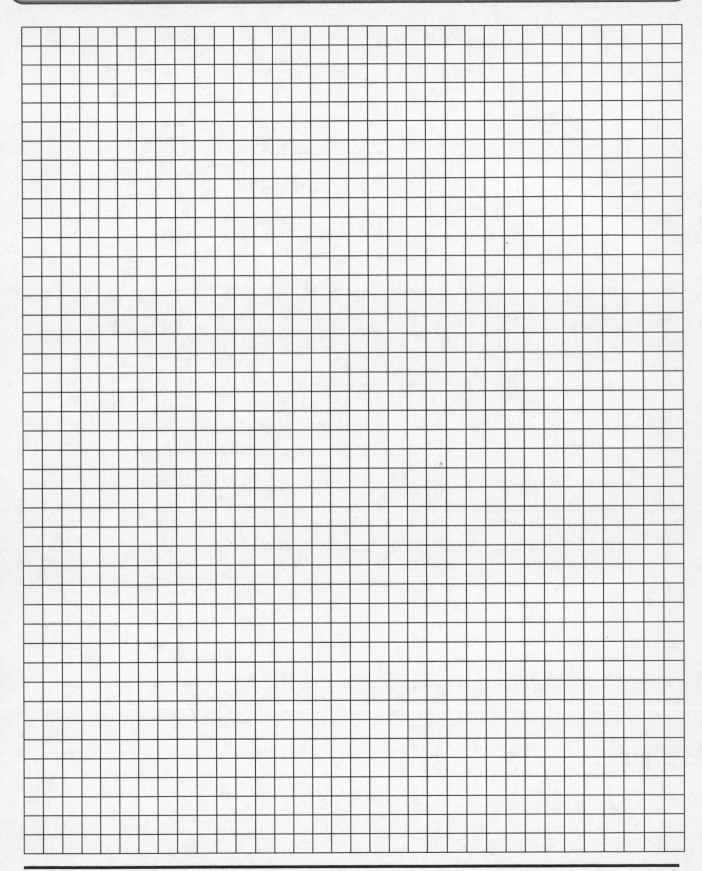

Notes